Polyimides

Edited by B.P. Nandeshwarappa
and Sandeep Chandrashekharappa

Published in London, United Kingdom

IntechOpen

Supporting open minds since 2005

Polyimides
http://dx.doi.org/10.5772/intechopen.94672
Edited by B.P. Nandeshwarappa and Sandeep Chandrashekharappa

Contributors
Pavitra Rajendran, Erumaipatty Rajagounder Nagarajan, Masao Tomikawa, Sudha V. Bhoraskar, Prashant S. S Alegaonkar, Vasant N. Bhoraskar, Guanming Yuan, Zhengwei Cui, B. P. Nandeshwarappa, Sandeep Chandrashekharappa, Manjunath S. Katagi, S.O. Sadashiv, Raghu Ningegowda, Sharangouda J. J. Patil, G. M. Shilpa

Notice
Statements and opinions expressed in the chapters are these of the individual contributors and not necessarily those of the editors or publisher. No responsibility is accepted for the accuracy of information contained in the published chapters. The publisher assumes no responsibility for any damage or injury to persons or property arising out of the use of any materials, instructions, methods or ideas contained in the book.

First published in London, United Kingdom, 2022 by IntechOpen
IntechOpen is the global imprint of INTECHOPEN LIMITED, registered in England and Wales, registration number: 11086078, 5 Princes Gate Court, London, SW7 2QJ, United Kingdom
Printed in Croatia

British Library Cataloguing-in-Publication Data
A catalogue record for this book is available from the British Library

Additional hard and PDF copies can be obtained from orders@intechopen.com

Polyimides
Edited by B.P. Nandeshwarappa and Sandeep Chandrashekharappa
p. cm.
Print ISBN 978-1-83969-884-2
Online ISBN 978-1-83969-885-9
eBook (PDF) ISBN 978-1-83969-886-6

We are IntechOpen,
the world's leading publisher of
Open Access books
Built by scientists, for scientists

6,000+
Open access books available

146,000+
International authors and editors

185M+
Downloads

Our authors are among the

156
Countries delivered to

Top 1%
most cited scientists

12.2%
Contributors from top 500 universities

Interested in publishing with us?
Contact book.department@intechopen.com

Numbers displayed above are based on latest data collected.
For more information visit www.intechopen.com

Meet the editors

B.P. Nandeshwarappa is an associate professor and chairman of the Department of Studies in Chemistry, Davangere University, India. He obtained an MSc and Ph.D. from Kuvempu University, India, and completed his postdoctoral studies at the Institute of Fine Chemicals, East China University of Science and Technology. He was previously a research associate at Akanksha Power and Infrastructure Pvt. Ltd. (APIPL), India. He has fourteen years of teaching experience and several textbooks, book chapters, and scholarly publications and presentations to his credit. He has guided more than fifty postgraduate dissertations and is currently supervising Ph.D. scholars. Dr. Nandeshwarappa has received several awards in recognition of his outstanding contributions. His research expertise and interests are centered on organic chemistry.

Dr. Sandeep Chandrashekharappa is an assistant professor in the Department of Medicinal Chemistry, National Institute of Pharmaceutical Education and Research, Raebareli (NIPER-R), India. He obtained an MSc and Ph.D. at Kuvempu University, India. Prior to his appointment at NIPER-R, Dr. Chandrashekharappa worked at the Institute for Stem Cell Science and Regenerative Medicine, India. He has more than thirteen years of experience in industrial and academic research. He has seventy-five journal articles and fifteen patents to his credit. He has presented several papers at national and international conferences. In recognition of scientific merit, he was awarded the prestigious Gandhian Young Technological Innovation award from the Government of India and the Best Researcher International Award. He has supervised graduate and postgraduate students. He is also a reviewer and editorial board member of several journals.

Contents

Preface

Polyimides are a new generation of polymers with inherently rigid chains that are of high commercial importance. This book discusses the properties, synthesis, and applications of polyimides, as well as presents future research directions and challenges. Polyimides possess excellent thermal, mechanical, and electrical properties and therefore have found many applications in broad technologies ranging from microelectronics to high-temperature matrices and adhesives for gas separation membranes. This book builds on the foundation of synthesis, characterization, and biological applications of polyimides. It places special emphasis on the synthesis and mechanism of the reactions of different classes of polyimides, such as polyimide films, which have been used in a variety of electrical and electronic insulation applications in industries such as aerospace, energy, automotive, and electronics. There has recently been an intense effort to develop flexible and transparent plastic substrates suitable for applications in conformable and roll-up displays. Polyimide films are potential candidates for these substrates owing to their excellent mechanical, thermal, chemical, and electrical properties. The ability of polyimide films to maintain their excellent properties over a wide temperature range has opened new areas of design and application. These materials are prepared by incorporating highly stable and rigid heterocyclic ring systems into the polymer chain. Their thermal stability is further significantly improved by incorporating aromatic rings on the backbone and side groups. This book provides a broad range of solutions to the problems of graded difficulties in polyimides.

Dr. B.P. Nandeshwarappa
Associate Professor,
DOS in Chemistry,
Davangere University,
Shivagangothri,
Karnataka, India

Dr. Sandeep Chandrashekharappa
Assistant Professor,
Department of Medicinal Chemistry,
National Institute of Pharmaceutical Education and Research,
Raebareli, (NIPER-R)

Department of Pharmaceuticals,
Ministry of Chemicals and fertilizers, Govt. of India,
Raebareli, Lucknow (UP), India

Chapter 1

Introductory Chapter: Polyimides - Importance and Its Applications

B.P. Nandeshwarappa, Sandeep Chandrashekharappa,
Manjunath S. Katagi, S.O. Sadashiv, G.M. Shilpa,
Raghu Ningegowda and Sharangouda J. Patil

1. Part A: introduction

Polyimides (PIs) are engineered polymers that resisted with high-temperature; these are exceptionally thermo stable with the combination of other ligands, observed mechanical toughness and chemical resistance. PIs exhibit excellent properties of dielectric and low coefficient the increasing condition of temperature. These are aromatic in nature belongs to the class of polymers and in the combination of ligands exhibit various properties. The major advantages of PIs are physical properties and mechanical characteristics in the higher scale of cryogenic temperatures (250–300°C) and heat resistance (400–450°C), factors that influence the set of properties with the presence of conjugated hetero aromatic compounds, also exhibit potent molecular interaction and strong chain bonds. Several methods are developed for the synthesis of PIs and all are familiar. The most familiar method for the synthesis of PIs is the two-stage and one-stage under thermo stable conditions using complex mixture of diamines and aromatic tetracarboxylic acids dianhydrides. The structures of these polymers (PIs) are exhibited as thermo labile plastic and provide very good chemical resistance, electrical, mechanical, and thermal properties. The application of PIs brings the following benefits to the medical tubing such as flexibility, high tensile strength, biocompatibility, low friction, transparency, tight tolerances, thin walls, smooth surface, push ability, and column strength. Polyimide was first developed at DuPont and reported in the year 1950s. In the last few years, PIs gained a lot of interest due to their huge applications as semi-conductor in electronic appliances. The superior properties of thermal, mechanical, and electrical PIs have made it possible use and various applications. PIs exhibit very lower amount of electrical leakage in applied surfaces. They act as inter layer, good dielectric insulators and also provide fine coverage in the fabrication stepwise in the multi layer form in IC structures. They are highly resisted with solvents and can be use it as sprayer, spun-on, traditional photo imaging as lithography and also engineered in etch process. PIs are a poly merofimide monomers belongs to the class of high performance synthetic plastics. Due to their high heat-resistance, they possess various applications and roles as harsh organic materials, like high temperature fuel cells, displays and unique features. A Kapton is a typical polyimide, which was synthesized by condensation reaction using pyromellitic dianhydride and 4,4′-oxydianiline materials [1]. The report of first polyimide

IntechOpen

discovery was made in 1908 by Bogart and Renshaw [2]. They found initially 4-amino phthalic anhydride which exhibit high heat along with water evaporation during the formation of polyimide which shows a high molecular weight. After that, the semi aliphatic polyimide was discovered by Edward and Robinson by the method of fusion using diamines and tetra acid as base materials [3]. The polyimide application was reported as commercially important as Kaptonin in the year 1950s by the workers at Dupont and they developed route for synthesis with the involvement of soluble polymer as a precursor. Till today similar method was used to synthesize as primary route to produce polyimides compounds. PIs have been started with this method for mass production in 1955. The fields, applications, types, composite mixtures, etc. of polyimides were extensively worked on and reported in many books, research articles [4–6], and review articles [7, 8]. Also found is literature on biobased polymers, to protect the environment, because petroleum-based materials are not eco-friendly, hence it is a matter of need of hour in the current societies [9]. Currently, many renewable resources are there such as proteins, carbohydrates and fats, these are employed to produce it as biodegradable material and alternative to the synthetic polyimides [10, 11]. In this context, packaging materials in industries mainly focused on the production of eco-friendly and work smartly, which includes antimicrobial protection, hydrophobicity, extension of shelf life, meet the demands of consumer and maximizing benefits [12]. These characteristics are significantly achieved by assisted techniques by using nanofillers, which act as supporting agents and nano carrier materials as antimicrobial substances [13]. Poly(L-lactic acid) (PLLA) and DL-lactide are produced on large scale in biological approach using of agricultural by-products in a microbial fermentation [14–17]. Owing to its beneficial features, exhibit very good mechanical properties, transparency, bio-degradability and bio-compatibility, hence it is considered best product industrial fields to produce it for wide range of application. It has been classified as a safe material by the Food and Drug Administration (FDA), due to that currently using in industries of food packaging applications [18]. As advanced methods PIs bounded with nanoparticles with different structures to develop more efficient and eco-friendly packaging materials to obtain excellent results. In the Santa Barbara Amorphous (SBA-15) and meso porous cellular foam (MCF) material as nanoparticles with 2D hexagonal shape along with cylindrical pores and 3D porous system in spherical structure were used to develop the PIs, respectively [19–21] and also they were act as nanocarriers for enhancing the chemical properties as preservatives to develop it into polymeric matrices with antimicrobial applications [22–24]. The advanced step as eco-friendly materials would be the greater extent to incorporation of natural key elements which exhibit useful properties by using essential oils [25–28]. However, their volatile nature may vary in the reaction, and it needs further characterization and optimization of the protocol [29, 30].

2. Part B: properties and applications

Thermo stable polyimides are familiar for thermal stability, strong chemical resistance, excellent mechanical support and its characteristic in orange/yellow color. Polyimides compounds affinity with graphite or glass fiber reinforcements have more flexural strengths measuring up to 340 MPa (49,000 psi) and flexural moduli of 21,000 MPa (3,000,000 psi). These properties makes them strong matrix of polyimides to exhibit low creep with extended tensile strength. Such properties are very vital when continuously use it to in higher temperatures like 232°C with minimizing excursions, even with the high temperature like 704°C. Molded products of polyimides and their laminates have a very good heat resistance.

Normal conditions of such products and laminates range exhibit cryogenic properties in which exceeding temperature at 260°C. Polyimides are naturally resistant against to flame combustion, they do not react it easily and usually exhibit as flame retardants. Generally polyimides are yellow color in nature. The polyimides which are synthesized and purified as aromatic dianhydrides are heavily colored. In the similar process, distilled and sublimed, *m*-and *p*-phenylenediamine are colorless crystalline solid materials, which exposed to air or in the liquid state, they produce colored characters, due to its oxidation reaction. Polyimides are radiant or opaque, the specific gravity of polyimides compounds are in the range of 1.1–1.5. Even though when polyimide compounds burn, they exhibit the inherent property of self-extinguishable to develop in to as surface charmaterial which smoothers the flame characters, hence it is preferred in the transportation sector and construction industries.

Author details

B.P. Nandeshwarappa[1*], Sandeep Chandrashekharappa[2], Manjunath S. Katagi[3], S.O. Sadashiv[4], G.M. Shilpa[5], Raghu Ningegowda[6] and Sharangouda J. Patil[7]

1 Department of Studies in Chemistry, Shivagangotri, Davangere University, Davanagere, Karnataka, India

2 Department of Medicinal Chemistry, National Institute of Pharmaceutical Education and Research (NIPER) Raebareli, Lucknow, UP, India

3 Department of Pharmaceutical Chemistry, Bapuji Pharmacy College, Davanagere, Karnataka, India

4 Department of PG Studies and Research in Food Technology, Shivagangothri Davangere University, Davanagere, Karnataka, India

5 Department of Physics, Sir MV Govt. Science College, Bommanakatte, Bhadravati, Karnataka, India

6 Department of Studies in Chemistry, Jyoti Nivas College Autonomous, Bengalore, India

7 Department of Zoology, NMKRV College for Women (Autonomous), Bengalore, India

*Address all correspondence to: belakatte@gmail.com

IntechOpen

References

[1] Jamshidian M, Tehrany EA, Imran M, Jacquot M, Desobry S. Poly-lactic acid: Production, applications nanocomposites, and release studies. Comprehensive Reviews in Food Science and Food Safety. 2010;**9**:552-571

[2] Lopez-Rubio A, Gavara R, Lagaron JM. Bioactive packaging: Turning foods into healthier foods throughbiomaterials. Trends in Food Science and Technology. 2006;**17**:567-575

[3] Alfei S, Schito AM, Zuccari G. Biodegradable and compostable shopping bags under investigation byFTIR spectroscopy. Applied Sciences. 2021;**11**:621

[4] Ahmed I, Lin H, Zou L, Brody AL, Li Z, Qazi IM, et al. A comprehensive review on theapplication of active packaging technologies to muscle foods. Food Control. 2017;**82**:163-178

[5] Alfei S, Marengo B, Zuccari G. Nanotechnology application in food packaging: A plethora of opportunitiesversus pending risks assessment and public concerns. Food Research International. 2020;**137**:109664

[6] Papageorgiou GZ, Terzopoulou Z, Bikiaris D, Triantafyllidis KS, Diamanti E, Gournis D, et al. Evaluation of the formed interface in biodegradable poly(l-lactic acid)/grapheme oxide nanocomposites and the effect of nanofillers on mechanical and thermal properties. Thermochimica Acta. 2014;**597**:48-57

[7] Gruber P, O'Brien M. Polylactides. Biopolymer. 2002:235-239

[8] Gupta AP, Kumar V. New emerging trends in synthetic biodegradable polymers—Polylactide: A critique. European Polymer Journal. 2007;**43**:4053-4074

[9] Hong CLS. An overview of the synthesis and synthetic mechanism of poly(lactic acid). Modeling Chemical Application. 2014;2-4

[10] Lasprilla AJR, Martinez GAR, Lunelli BH, Jardini AL, Filho RM. Poly-lactic acid synthesis forapplication in biomedical devices—A review. Biotechnology Advances. 2012;**30**:321-328

[11] Vert M, Schwarch G, Coudane J. Present and future of PLA polymers. Journal of Macromolecule Science. 1995;**32**:787-796

[12] Ataman-Önal Y, Munier S, Ganée A, Terrat C, Durand PY, Battail N, et al. Surfactant-free anionic PLA nanoparticles coated with HIV-1 p24 proteininduced enhanced cellular and humoral immune responses in various animal models. Journal of Controlled Release. 2006;**112**:175-185

[13] Shen W, Zhang G, Li YL, Fan G. Effects of the glycerophosphate-polylactic copolymer formation onelectrospun fibers. Applied Surface Science. 2018;**443**:236-243

[14] Conn RE, Kolstad JJ, Borzelleca JF, Dixler DS, Filer LJ, Ladu BN, et al. Safety assessment ofpolylactide (PLA) for use as a food-contact polymer. Food and Chemical Toxicology. 1995;**33**:273-283

[15] Siafaka PI, Barmbalexis P, Bikiaris DN. Novel electrospunnanofibrous matrices prepared from poly(lacticacid)/poly(butylene adipate) blends for controlled release formulations of an anti-rheumatoid agent. European Journal of Pharmaceutical Science. 2016;**88**:12-25

[16] Freiberg S, Zhu XX. Polymer microspheres for controlled drug

release. International Journal of Pharmaceutics. 2004;**282**:1-18

[17] Park S-Y, Pendleton P. Mesoporous silica SBA-15 for natural antimicrobial delivery. Powder Technology. 2012;**223**:77-82

[18] Farjadian F, Roointan A, Mohammadi-Samani S, Hosseini M. Mesoporous silica nanoparticles: Synthesis,pharmaceutical applications, biodistribution, and biosafety assessment. Chemical Engineering Journal. 2019;**359**:684-705

[19] Wang Y, Zhao Q, Han N, Bai L, Li J, Liu J, et al. Mesoporoussilica nanoparticles in drug delivery and biomedical applications. Nanomedical and Nanotechnology Biology Medicine. 2015;**11**:313-327

[20] Nanaki S, Siafaka PI, Zachariadou D, Nerantzaki M, Giliopoulos DJ, Triantafyllidis KS, et al. PLGA/SBA-15 mesoporous silica composite microparticles loaded with paclitaxelfor local chemotherapy. European Journal of Pharmaceutical Sciences. 2017;**99**:32-44

[21] Vallet-Regi M, Rámila A, Del Real RP, Pérez-Pariente J. A new property of MCM 41: Drug deliverysystem. Chemistry of Materials. 2001;**13**:308-311

[22] Gao F, Zhou H, Shen Z, Qiu H, Hao L, Chen H, et al. Synergistic antimicrobial activities of teatree oil loaded on mesoporous silica encapsulated by polyethyleneimine. Journal of Dispersion Science and Technology. 2020;**41**:1859-1871

[23] Jin L, Teng J, Hu L, Lan X, Xu Y, Sheng J, et al. Pepper fragrant essential oil (PFEO)and functionalized MCM-41 nanoparticles: Formation, characterization, and bactericidal activity. Journal of Science and Food Agriculture. 2019;**99**:5168-5175

[24] Nanaki S, Tseklima M, Terzopoulou Z, Nerantzaki M, Giliopoulos DJ, Triantafyllidis K, et al. Use of mesoporous cellular foam (MCF) in preparation of polymeric microspheres for longacting injectable release formulations of paliperidone antipsychotic drug. European Journal of Pharmaceutics and Biopharmaceutics. 2017;**117**:77-90

[25] Nazzaro F, Fratianni F, De Martino L, Coppola R, De Feo V. Effect of essential oils on pathogenic bacteria. Pharmaceuticals. 2013;**6**:1451-1474

[26] Devi KP, Nisha SA, Sakthivel R, Pandian SK. Eugenol (an essential oil of clove) acts as an antibacterial agentagainst salmonella typhi by disrupting the cellular membrane. Journal of Ethnopharmacology. 2010;**130**:107-115

[27] Ben Arfa A, Combes S, Preziosi-Belloy L, Gontard N, Chalier P. Antimicrobial activity of carvacrolrelated to its chemical structure. Letters in Applied Microbiology. 2006;**43**:149-154

[28] Raut JS, Karuppayil SM. A status review on the medicinal properties of essential oils. Industrial Crops and Products. 2014;**62**:250-264

[29] AlkanTas B, Sehit E, ErdincTas C, Unal S, Cebeci FC, Menceloglu YZ, et al. Carvacrol loaded halloysitecoatings for antimicrobial food packaging applications. Food Packaging and Shelf Life. 2019;**20**:100300

[30] Melendez-Rodriguez B, Figueroa-Lopez KJ, Bernardos A, Martínez-Máñez R, Cabedo L, Torres-Giner S, et al. Electrospun antimicrobial films of poly(3-hydroxybutyrate-co-3-hydroxyvalerate) containing eugenol essential oil encapsulated in mesoporous silica nanoparticles. Nanomaterials. 2019;**9**:227

Chapter 2

Polyimide-Derived Graphite Films with High Thermal Conductivity

Guanming Yuan and Zhengwei Cui

Abstract

Nowadays, polyimide-derived graphite films with high thermal conductivity have been increasingly applied in many cutting-edge fields needing thermal management, such as highly integrated microelectronics and wireless communication technologies. This chapter first introduces a variety of functional graphite films with high thermal conductivity of 500–2000 W/m K in the planar direction, then provides the preparation technology (including lab-scale preparation and industrial production) and quality control strategy of high-thermal-conductivity graphite films, which are derived from a special polymer- polyimide (PI) by carbonization and graphitization treatments through a suitable molding press in a vacuum furnace. The morphology, microstructure and physical properties as well as the microstructural evolution and transformation mechanism of PI films during the whole process of high-temperature treatment are comprehensively introduced. The nature of PI precursor (e.g., the molecular structure and planar molecular orientation) and preparation technics (e.g., heat-treatment temperature and molding pressure) are critical factors influencing their final physical properties. Currently challenged by the emerging of graphene-based graphite films, the latest developments and future prospects of various PI-derived carbons and composites (beyond films) with high thermal conductivity have been summarized at the end. This chapter may shed light on a promising and versatile utilization of PI-derived functional carbon materials for advanced thermal management.

Keywords: polyimide, graphite film, preparation, structure, property, high thermal conductivity, thermal management

1. Introduction

Highly oriented graphite film has excellent electrical and thermal conductivity properties, and is an ideal material indispensable for the development of modern science and technology. It has a very broad application prospect in thermal management field such as modern microelectronic packaging-integration and 5 G wireless communication technologies. In the early 1960s, scientists had used high-temperature pyrolysis deposition technology to prepare highly oriented pyrolytic graphite (HOPG), however, the material needs to be prepared at high temperature (up to 3400-3600°C) and high pressure (10 MPa), the production cycle is long and the production cost is high. Thus, the wide application of such material is subject to certain restrictions [1]. Subsequently, Japanese scientists had initially discovered that polyimide (PI) film with a golden appearance as shown in **Figure 1a** did not melt during the carbonization process and maintained the original film shape,

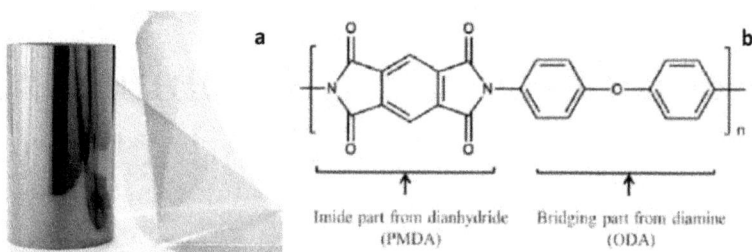

Figure 1.
(a) Optical appearance and (b) a molecular repeating unit of Kapton PI film [2].

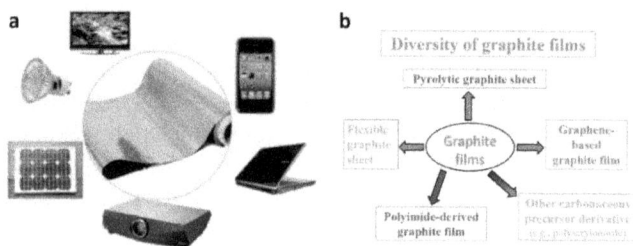

Figure 2.
(a) Wide heat-dissipation application of PI-derived graphite films in advanced microelectronics and (b) a variety of graphite films classified by different precursors.

after high-temperature (2800-3200°C) graphitization treatment, a highly oriented graphite film with a structure close to single crystal graphite can be obtained [3, 4]. Nowadays, PI developed as a thermoresistant polymer has been widely used in different fields, for instance, aromatic PI is often employed as an excellent carbonaceous precursor to prepare various carbon materials with different morphologies (e.g., fiber, film, foam and block) [2]. This is because aromatic PI has many advantages such as wide range of well-defined molecular structure as shown in **Figure 1b**, relatively high crystallinity and carbon yield.

Recently, PI-derived graphite films with high thermal conductivity in the planar direction ranging in 500–1900 W/m K have been successfully produced and practically applied in heat dissipation of many microelectronic devices as shown in **Figure 2a**. This is attributed to the extensive research on the composition, structure and properties of PI polymer film and related high-temperature heat treatment process have been conducted to improve the thermal conductivity of resultant graphite films and reduce the production cost [2–13]. It is well-known that the thermal conductivity of graphite films is greatly affected by many factors (the quality of PI film precursor, film thickness and heat treatment temperature, etc.). In addition, the microstructural evolution and transformation mechanism [4, 6, 12] of PI polymer during high-temperature heat treatment, the capability of forming an ordered graphite structure and the relevant control strategy need to be further understood. This will make the application of graphite films for thermal management move forward [2, 4, 11].

2. Diverse graphite films with high thermal conductivity

Generally, high-thermal-conductivity graphite films can be divided into two main types (natural graphite-derived and artificial synthetic films) and several subdivided categories as shown in **Figure 2b** according to different raw materials: oriented pyrolytic graphite sheet, flexible graphite sheet, graphene-based graphite film, PI-derived graphite film and other carbonaceous precursor derivative.

HOPG sheet refers to polycrystalline graphite film with a high bulk density of ~2.20 g/cm^3 and highly oriented graphene layers stacking along the c-axis direction, similar to single crystal graphite as shown in **Figure 3** [14]. Its room-temperature thermal conductivity along the a-axis direction of the graphite sheet reaches up to 1600–2000 W/m K [15]. Recently, through a facile and feasible chemical vapor deposition on transition metal substrates, the prepared graphite films possess a high thermal conductivity of 600–1570 W/m K [16, 17].

Flexible graphite sheet is prepared by using natural flake graphite as raw material through several procedures as follows. Firstly, strong acidification treatment for chemical intercalation, then washing, drying and high-temperature expansion to obtain high-expanded graphite worms, and finally calendering and pressing treatment processes. The thermal conductivity of flexible graphite sheet can be adjusted in the range of 200–600 W/m K according to the bulk density and sheet thickness [18, 19]. Because this material does not require high-temperature graphitization and the preparation process is simple, the production cost is relatively low, and it can be used not only as a high-temperature sealing material, but also as a heat dissipation pad for the interface between electronic devices and heat sinks. In addition, the thin graphite sheet has a certain degree of flexibility and can be bent and rolled for storage as shown in **Figure 4**, which accelerates its low-cost industrial production. However, the mechanical properties of flexible graphite sheet decrease with the increase of thickness. So it is suitable for fields where material strength, toughness and flexural properties are not very high.

Figure 3.
(a) Electron channeling pattern and (b) SEM image of HOPG [14].

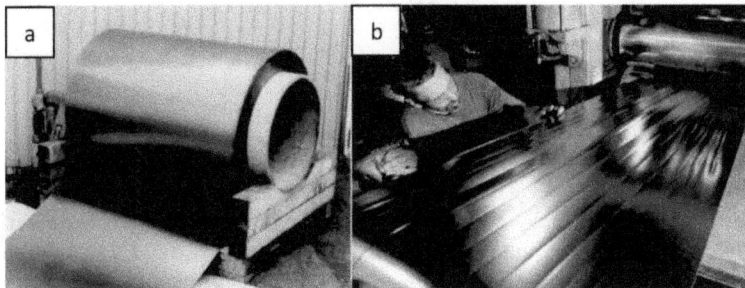

Figure 4.
Optical photographs of (a) flexible graphite sheet rolled for store in Nihon carbon and (b) GrafTech graphite sheet production line.

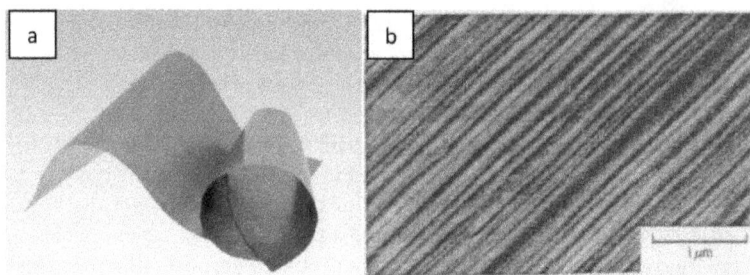

Figure 5.
(a) Optical photograph and (b) microscopic image of PI-derived thermal pyrolytic graphite sheet (PGS) produced by Panasonic Industry [20].

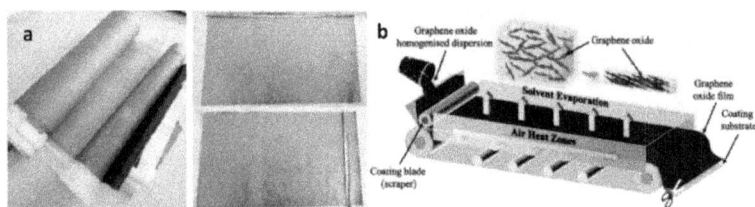

Figure 6.
(a) Optical photograph of large-scale preparation of graphene laminated films [21] and (b) the schematic of film fabrication process [22].

PI graphite film with high crystallinity and preferred orientation is similar to HOPG as shown in **Figure 5** and has a high thermal conductivity up to 1900 W/m K in the planar direction [2]. The thermal conductivity of the pyrolytic graphite sheets developed by Panasonic Industry is reported to be 700-1950 W/m K according to their bulk densities (0.85–2.13 g/cm^3) and sheet thicknesses (0.10–0.01 mm) [20]. Obviously, the thermal conductivity of graphite sheets is greatly affected by its bulk density, thickness and pyrolysis process. As a rule, the greater the thickness is, the lower the thermal conductivity is.

Recently, there has been numerous studies on making graphene or its precursor (e.g., graphene oxide and reduced graphene oxide) into graphite films as shown in **Figure 6a** [21–23]. This extensive research greatly improves the thermal conductivity of graphene-based films up to 2000 W/m K and promotes their various applications [23]. Although the production cost of large-area high-quality graphene films with high thermal conductivity is still high at this stage, the large-scale fabrication in science and industry is rising [21, 22].

3. Preparation technology of graphite films

3.1 Lab-scale preparation

At present, a batch-scale preparation method, i.e., multi-sheet carbonization by laminated molding in a small-sized vacuum induction furnace as shown in **Figure 7a**, is generally adopted to prepare sheet-like PI-derived graphite films with different sizes. However, this method shows obvious disadvantages such as relatively high production cost, low preparation efficiency, and particularly large energy consumption. Moreover, the size of the films is limited by the graphite

Figure 7.
Optical photographs of (a) a lab-scale vacuum induction furnace and (b) roll-shaped graphite film prepared from PI film.

mold (which needs to repeatedly endure severe condition under high temperature and high pressure) and the size of the heat treatment furnace.

3.2 Industrial production

In general, the industrial production of PI-derived graphite films is carried out in multiple sets of large-sized vacuum induction furnace. The emerging advanced rolling-carbonization technology, i.e., directly performing carbonization-graphitization treatment on the roll-shaped organic PI films, could obtain a roll-shaped graphite films with a large size as shown in **Figure 7b** by controlling the heat treatment process of tightly rolled films. This technology can significantly improve the production efficiency of graphite films, reduce the production cost and improve their mechanical properties.

3.3 New fabrication technology

With the rapid development of graphene and graphene-based materials throughout the world, some new methods, new processes and new technologies [21, 22, 24–27], such as molecular welding, molecular assembling, flow coating and centrifugal casting, as shown in **Figures 6b** and **8**, have been increasingly developed to fabricate graphene-based graphite films with high thermal conductivity for thermal management application. This will surely provide some reference for the preparation strategy of PI-derived graphite films. It is possible to take advantage of continuous high-temperature carbonization technology as shown in **Figure 8b** to fabricate large-scale PI-derived graphite films at a low cost in the future.

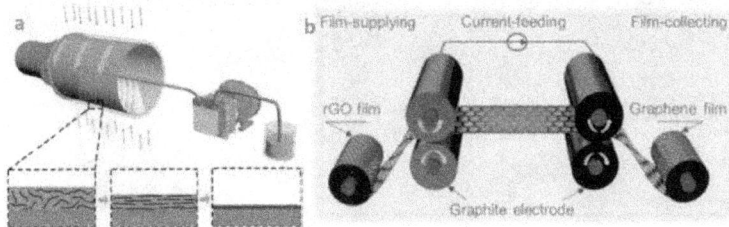

Figure 8.
Schematics of producing graphene-derived films by different methods (a) the continuous centrifugal casting [26] and (b) the pressurized roll-based production by Joule heating [27].

4. Morphology, structure and properties of PI-derived graphite films

The uniform PI raw film (DuPont Kapton) with a golden color shrinks significantly after 1000°C carbonization and 3000°C graphitization under proper pressure in a vacuum furnace, and the shrinkage rate in the planar direction is about 15% × 15%. The color of the film changes from yellow to black and gray as shown in **Figure 9a–c**, the carbonized and graphitized samples are brittle and flexible (can be bent at a certain angle >90° at many times), respectively. The molding-press and its pressing strength on the PI films have an important role on the final quality of resultant graphite films as shown in **Figure 9d–f**.

As shown in **Figure 10**, the surface of 1000°C-carbonized carbon films (with a thickness of 50 μm for the raw film) is smooth and the thickness is still uniform, the internals of carbon films exhibit an amorphous carbon structure. After 2000°C graphitization, a local chaotic layered structure can be observed in the cross-section of the films [28]. When the graphitization temperature reaches 2400°C, the cross-section of the films presents a more uniformly oriented layered structure, and as the graphitization temperature further increases up to 3000°C, the layered structure becomes more flatted and ordered, and the graphite-like crystal structure is nearly perfect.

The thickness and the nature of the PI films have a significant impact on the capability of forming a graphite-like crystal structure. As shown in **Figure 11a–c**, the PI film with a thickness of 50 μm completely forms a graphite-like layered structure with high crystallinity, and the degree of preferred orientation of the graphene layers is high. The PI films with thicknesses of 75 and 100 μm display a partial graphite-like layered structure and nearly amorphous structure with low crystallinity and poor crystalline orientation, respectively. Some small holes appear on the cross-section of graphite film, which may be related to the removal of non-carbon elements during the high-temperature graphitization process. Moreover, the nature of PI films (e.g., the variety of polymer constituent and molecular structure

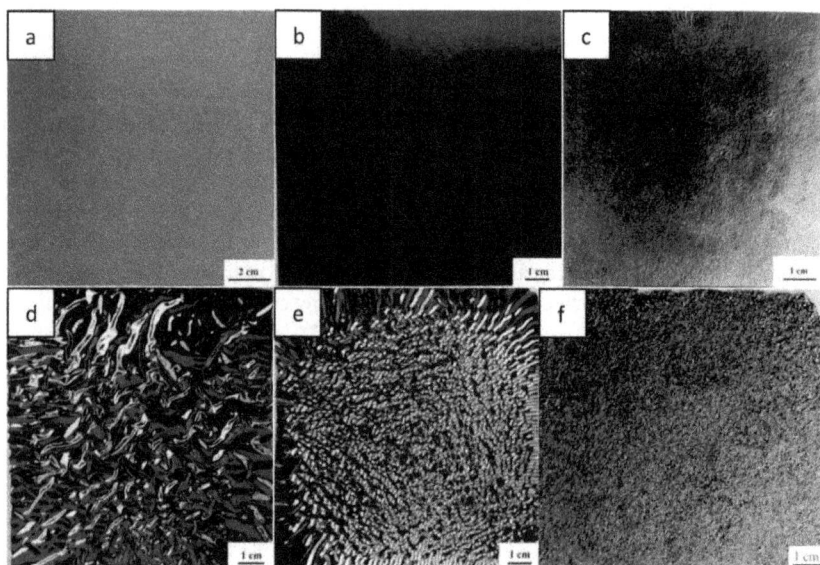

Figure 9.
Optical photographs of (a) PI raw film, its (b, d and e) 1000°C-carbonized and (c and f) 3000°C-graphitizatized samples made by various molding-press treatments ((b and c) suitable pressure; (d) no pressure; (e) insufficient pressure; and (f) excessive pressure).

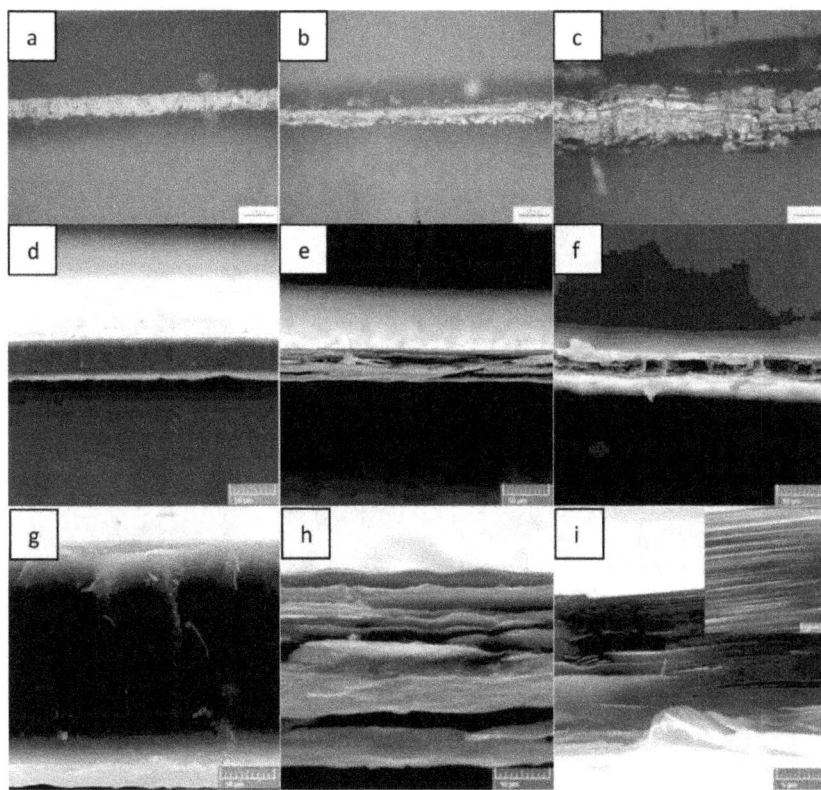

Figure 10.
(a–c) PLM and (d–i) SEM images of the transversal section of PI films heat-treated at different temperatures ((a, d and g) 1000; (b, e and h) 2400; and (c, f and i) 3000°C), (g–i) are high magnification images of (d–f), respectively, and the top right inset in i is the corresponding enlargement.

in various PI films produced by different manufacturers) is very critical to prepare highly oriented graphite films, which has been demonstrated in **Figure 11d–f**.

The PI laminated sample with a good graphite-like crystal structure as shown in **Figure 12** could be prepared by a hot-press method at 2400°C under a certain pressure. It has a uniformly layered structure in the cross-section, and the stacking of PI films is regular and orderly. The PI monolayer film inside the laminated sample still maintains its complete sheet-like structure, which is conducive to the high efficient conduction of heat in the two-dimensional direction of the plane. The PI-derived laminates can be used as a bulk thermally conductive material to further expand the application field of graphite films, but the controllable preparation of such large-size and ultra-thick bulk materials (e.g., blocks) is still difficult [4].

It can be seen from the XRD patterns as shown in **Figure 13a–c** that the PI raw film has a certain degree of orientation owing to the arrangement of aromatic molecules. With the rise of heat treatment temperature, the intensity of diffraction peak of the (002) crystal plane of the PI sample continues to increase. Meanwhile, the microcrystallite accumulation height (Lc) and graphitization degree (g) listed in **Table 1** increase step by step. After 3000°C graphitization, the interlayer spacing d_{002} (0.336 nm) is close to the theoretical value of single crystal graphite (0.3354 nm). The ratio of the two peaks (D and G) as shown in **Figure 13d** gradually decreases, especially the D peak of 2400°C-graphitized sample completely disappears, which indicates that a three-dimensional ordered graphite structure forms in the graphite film, the content of amorphous carbon and structural defects

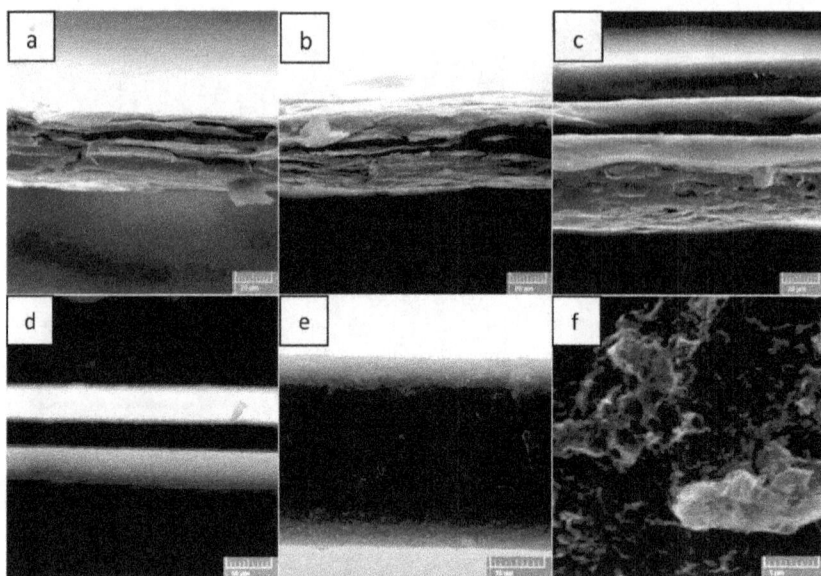

Figure 11.
SEM images of the transversal section of 3000°C-graphitized films derived from Kapton PI films with different thicknesses of ((a) 50, (b) 75, (c) 100 μm) and other brand PI film with a thickness of 50 μm at different enlargements (d–f).

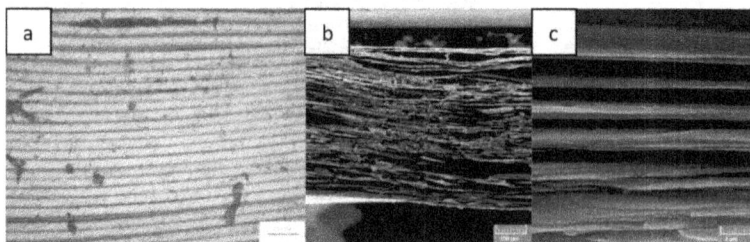

Figure 12.
(a) PLM and (b and c) SEM images of the transversal section of PI film-stacked block made by a suitable molding-press treatment at 2400°C.

is very low, and the graphite crystalline size is large [29]. It is worth noting that the microcrystalline size and g of graphite films are affected significantly by the nature (e.g., the extent of biaxial stretching on the original film) and thickness of PI films. The microcrystals in the thick graphite films grow and crystallize slowly, and their preferred orientation is relatively low. As a comparison, the graphite films made from other brand PI show an amorphous structure after graphitization at 3000°C, their microcrystals are small and disordered. The higher the heat-treatment temperature is, the easier the structural transformation completes. Graphitization treatment results in the better growth and crystallization of graphite microcrystals and the preferable orientation of graphene layers in the graphite films.

Figure 14a shows the room-temperature electrical resistivities of the Kapton PI films after heat treatment at different temperatures. It can be seen that the electrical resistivities of the PI films decrease significantly with the increase of the heat treatment temperature, indicating that the electrical conductivities increase rapidly. The PI film is a polymer insulating material and its volume electrical resistivity is as high as 10^{16} Ω cm. After 1000°C carbonization treatment, the electrical resistivity

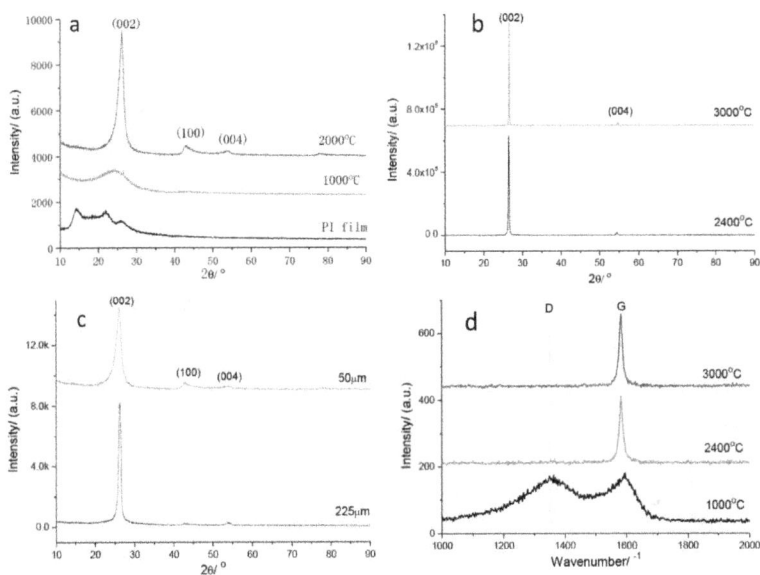

Figure 13.
(a–c) XRD patterns and (d) Raman spectra of various PI films ((a, b and d) Kapton; (c) other brand) heat-treated at different temperatures.

Sample	$2\theta_{002}/°$	d_{002}/nm	Lc/nm	g/%
PI raw film	25.94	0.343	2.03	10
PI-1000°C-50 μm	24.37	0.365	3.06	—
PI-2000°C-50 μm	26.12	0.341	5.63	34
PI-2400°C-50 μm	26.44	0.338	39.83	70
PI-2800°C-50 μm	26.50	0.337	50.95	82
PI-3000°C-50 μm	26.56	0.336	65.94	93
PI-3000°C-100 μm	26.33	0.338	49.71	70
PI-3000°C-50 μm[a]	26.22	0.340	12.42	47
PI-3000°C-225μm[a]	26.01	0.342	5.63	23
Single crystal graphite	26.58	0.3354	>100	100

[a]*Other brand PI film.*

Table 1.
Microcrystalline parameters of various PI films heat-treated at different temperatures.

reduces by 18 orders of magnitude, to about 54.6 μΩ m, because the PI film has undergone structural changes at this time, most of the heteroatoms are eliminated, and the carbon content increases significantly. At this stage, a local hexagonal-like carbon layer structure forms in the interior of carbon film. The electrical resistivities of the graphitized samples at 2000 and 2800°C are 5.5 and 0.82 μΩ m, respectively. The decline is not very large due to the fact that the conductive path in PI film has been formed around 2000°C. Further graphitization is only to improve its three-dimensional ordered structure with highly preferred orientation as shown in **Figure 10**. The electrical resistivity of the 3000°C-graphitized PI film is as low as 0.48 μΩ m, which is very close to the theoretical electrical resistivity of single crystal graphite (0.4 μΩ m) in the planar direction [30]. With the rise of heat treatment

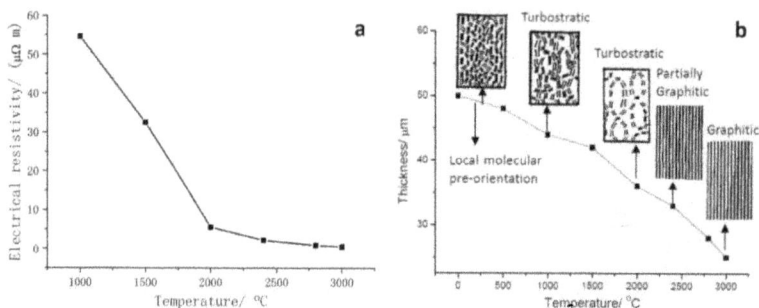

Figure 14.
(a) Room-temperature electrical resistivities of PI films heat-treated at different temperatures in the planar direction and (b) micro-structural evolution and transformation mechanism model from PI polymer to ordered graphite film during high-temperature treatment reproduced from [6].

temperature, the g of PI films continues to increase as listed in **Table 1**, and its internal graphene layered structure with highly preferred orientation is conducive to the transmission of electrons [2, 4].

From the above discussion on the morphology and microstructure of the PI films heat-treated at different temperatures, a microstructural change model from PI polymer to ordered graphite at each stage is shown in **Figure 14b** [6]. The heat-treatment process can be roughly divided into four stages: the first stage (500-1000°C), the second stage (1000-2000°C), the third stage (2000-2400°C) and the fourth stage (2400-3000°C). The whole process reflects that the internal structure of the PI film gradually changes from a disorderly amorphous structure to a highly crystalline graphite structure as the heat treatment temperature progresses [4].

According to the relevant empirical formulas [31], the thermal conductivity of 3000°C-graphitized graphite films (with a thickness of ca. 25 μm) is calculated to be 1143 W/m K. Measured by a laser thermal conductivity meter (NETZSCH LFA 457), its room-temperature thermal diffusion coefficient is ~700 mm^2/s, and the corresponding thermal conductivity is measured to be 994 W/m K (the bulk density and specific heat are about 2.0 g/cm^3 and 0.71 J/g K, respectively). This excellent conduction performance is attributed to the highly ordered three-dimensional graphite structure of this film material.

5. Influencing factors on thermal conductivity of graphite films

It is well-known to all that the high-thermal-conductivity of carbon materials comes from the strong C—C covalent bonding between carbon atoms and the highly ordered graphite structure stacked by graphene layers and mainly results from the anharmonic vibration of the elastic lattice (i.e., the mutual interaction of phonons) to transfer heat [32]. Single crystal graphite has a hexagonal network layered structure and an anisotropic thermal conductivity, as shown in **Figure 15a**, its thermal conductivity along the a-axis direction (as high as 2000 W/m K) is much greater than that along the c-axis direction [33]. However, for carbon materials with a disordered graphite structure, the graphene layers with different sizes are stacked randomly, a lower thermal conductivity will yield unexpectedly. There are many critical factors governing the heat-dissipation performance of graphene-assembling carbon materials, such as microcrystalline size, crystalline orientation, structural defects (e.g., vacancies and substitution) and wrinkle deformation in graphene layers as shown in **Figure 15b** [34].

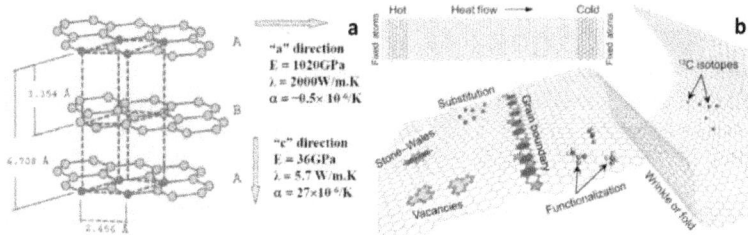

Figure 15.
(a) Crystal structure of perfect graphite with anisotropic thermal conductivity reproduced from [33] and (b) key factors determining heat dissipation of graphene-assembling film materials [34].

Usually, organic carbonaceous compounds are used as raw materials to prepare carbon materials. Under low temperature at about 300-1000°C, the component containing H, O, N and other non-C elements in organic compounds is gradually decomposed, and C-containing aromatic molecules continue to cyclize and aromatize, which forms C-rich material (i.e., carbon material), and finally through the graphitization process up to 3000°C, pure C material, i.e., graphite material can be obtained. Most of the chemical reactions during the carbonization of precursors are accompanied by the evolution of various gases—different hydrocarbons, carbon oxides, and H_2 [35]. It is important to timely remove the pyrolytic gases from the stress-stacked PI films in the highly sealed furnace. The conversion from PI polymer film to graphite film is a typical process of solid phase carbonization. Its prominent characteristic is the similarity in morphology (and shape) of raw material and final product without experiencing a fusion process, which is different from that of liquid phase carbonization [36]. Therefore, selecting proper carbonaceous precursors (e.g. Kapton PI film) and appropriate heat treatment process (e.g., high temperature graphitization under a suitable pressure and duly degassing treatment) to control the growth, accumulation and orientation of graphite microcrystals inside the carbon materials as shown in **Figure 16a** [37], are essential for obtaining graphite films with high thermal conductivity.

As a result, the thermal conductivity of graphite films mainly depends on the nature of the polymer films and their capability of forming an ordered graphite structure through high-temperature heat treatment as diagramed in **Figure 16b**. There are three mainly important conditions for obtaining graphite films with high thermal conductivity as follows [2, 4, 7, 28]. Firstly, high carbon content in the large

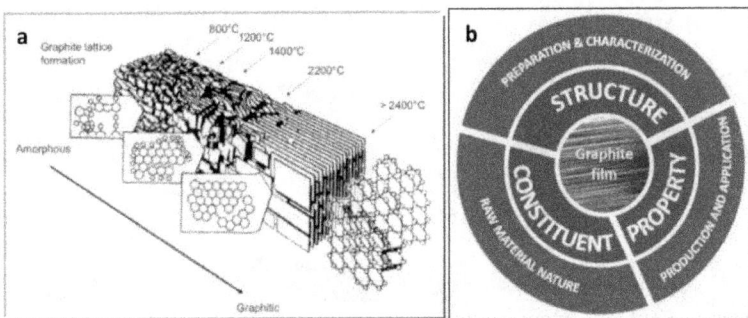

Figure 16.
(a) Marsh-Griffiths model of carbonization-graphitization process on a carbonaceous precursor [37] and (b) a diagram of quality control strategy for high-thermal-conductivity graphite films.

molecules and high carbon yield after carbonization treatment. Secondly, the quality of polymer films (e.g., the constituents and structure of aromatic molecules, high molecular planarity and suitable stiffness as well as molecular orientation degree through the role of biaxial stretching treatment, appropriate film thickness). Thirdly, heat treatment process control (e.g., heating treatment procedure, molding-press condition, non-carbon elements escape, and final graphitization temperature).

6. Latest developments of PI-derived carbons with high thermal conductivity

6.1 Modification of PI precursor to improve the flexibility of graphite films

There is no denying that the PI-derived graphite films with high thermal conductivity after graphitization treatment have a certain degree of brittleness as shown in **Figure 17a** due to their high stiffness (modulus) and high crystallinity and crystalline orientation, which undoubtedly limits their wide applications. It is difficult to achieve high thermal conductivity and ideal mechanical properties for the graphite films (e.g., internally contradictory indices like high modulus (associating with thermal conductivity) and high flexibility are hardly satisfied simultaneously) except a few reports such as Refs [2, 11, 39, 40]. Nowadays, the modification by doping of PI precursor with graphene (and graphene oxide) and other precursors (e.g., polyacrylonitrile) is a good strategy to improve the flexibility of graphite films with high thermal conductivity [38, 41–43] as shown in **Figure 17b**. It is interesting to note that various striking cranes with good flexibility as shown in **Figure 18** made with different raw materials by different methods and processes have been successfully prepared [2, 19, 38, 43].

6.2 Versatile forms of PI-derived carbons

Recently, many new forms of PI-derived carbons (including carbon fibers, carbon foams, carbon aerogels and carbon blocks as shown in **Figure 19**, which are beyond graphite films) with a feature of high thermal conductivity have been fabricated [44–51]. This extensive and intensive research on PI polymer will expand its application areas. Especially, ultrathick graphene film with a high thickness up to 200 μm while retaining a high thermal conductivity of 1200 W/m K has been achieved [52], which will stimulate the preparation of ultrathick (e.g., millimeter-scale) PI-derived graphite films or large graphite blocks.

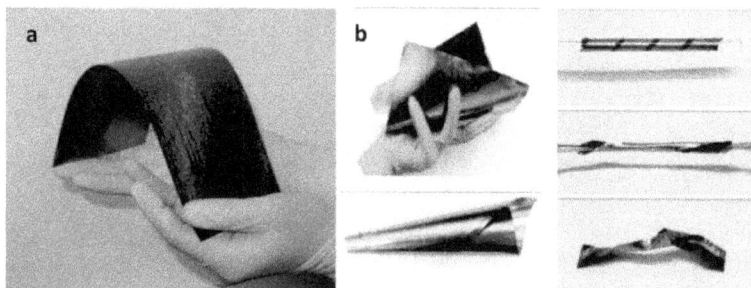

Figure 17.
Optical photographs of (a) highly-oriented PI-derived graphite thin sheet with improved manual handling [11] and (b) superflexible (bending, curling, enwinding, twisting, and knotting) graphene films [38].

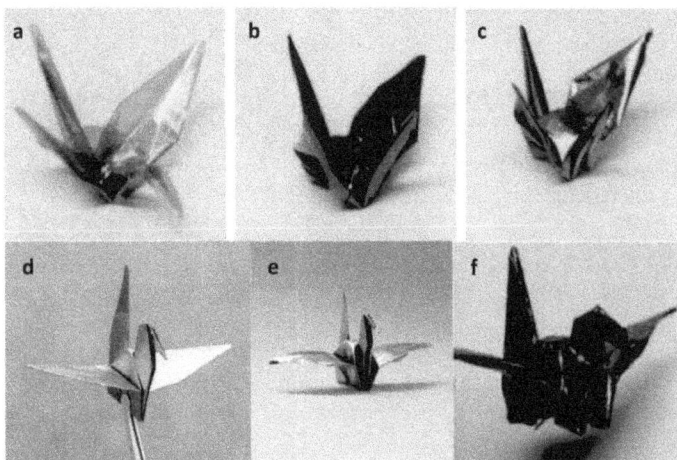

Figure 18.
Optical photographs of evolutional crane made of a PI film (a) and after carbonization (b) and graphitization (c) treatments showing good shape-retention and flexibility [2] and (b) various cranes derived from superfoldable graphene film (d) [38], polyacrylonitrile-derived graphite film (e) [43] and flexible graphite sheet (f) [19].

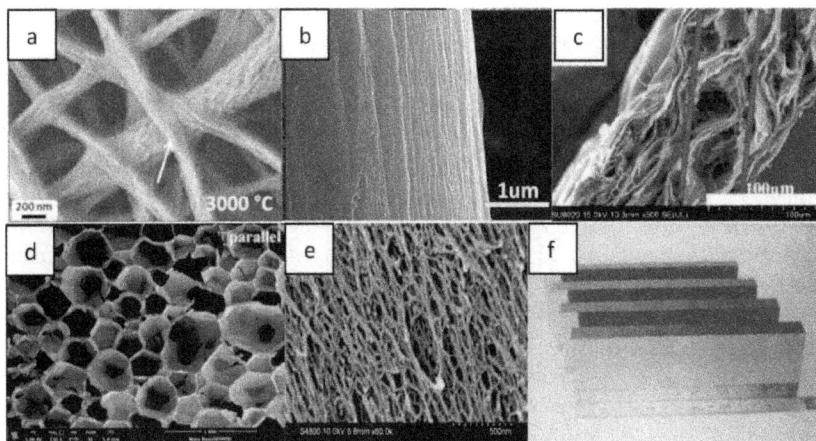

Figure 19.
(a–e) SEM images of various PI-derived carbons with high thermal conductivity ((a) carbon nanofibers [44], (b) carbon microfibers [45], (c) carbon bubbles [46], (d) carbon foams [47], (e) carbon aerogels [48]) and (f) optical photograph of highly oriented graphite blocks prepared from PI [49].

6.3 PI-derived composite materials

It is well accepted that graphene-based carbon films as thermal management materials can boost the heat-dissipation performance of film materials in the planar direction [21–23, 32–34, 38–41, 53, 54]. Through new functional composite technologies (e.g., chemical interaction as shown in **Figure 20a** [53], and modification treatment through doping or hybridizing with other carbonaceous precursors (graphene, carbon nanotube, etc.) and non-carbon fillers such as BN) [53–56], the thermal conductivity and mechanical flexibility of resultant graphite films can be both enhanced. Furthermore, in the through-plane direction, a superhigh thermal conductivity up to 150 W/m K can be obtained by novel structure design as shown in **Figure 20b** [54]. This affords carbon materials with a feature of three-dimensional

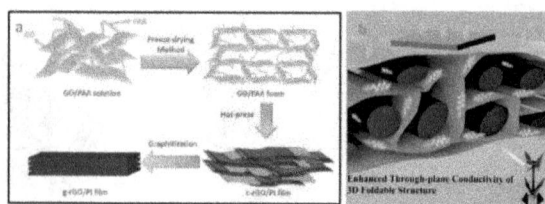

Figure 20.
(a) A diagram of preparing composite films through the covalent bonding between graphene oxide and PI [53] and (b) a 3-D hybridized structure of composite carbon film made with graphene oxide and PI [54].

high thermal conductivity (it is beyond the traditional graphite materials with high thermal conductivity only in the planar direction [32]), which will further promote the wide practical applications of carbon materials in thermal management [22, 23, 53–55, 57–59].

7. Conclusions

In a lab-scale study, the Kapton PI-derived graphite films (with a thickness of 50 μm for raw film) show a three-dimensional ordered structure consisting of graphene layers with highly preferred orientation and prefect graphite crystals after graphitization at 3000°C. Their electrical resistivity and thermal conductivity at room temperature in the planar direction are 0.48 μΩ m and ~ 1000 W/m K, respectively. The nature of PI precursor (the molecular structure, planar molecular orientation and film thickness, etc.) and preparation technics (e.g., heat-treatment temperature and molding pressure) have a critical influence on the final conduction performance of graphite films.

In the early time, limited by preparation technology, the thickness of PI-derived graphite films were mainly 20–50 μm, and their thermal conductivity in the planar direction was mostly 300–1000 W/m K. With the continuous improvement of production technology, high-thermal-conductivity graphite film products become more abundant, and some are even industrialized. The 10 μm-thin graphite films can approach a high thermal conductivity of 1900 W/m K. Currently, the thickest product (derived from graphene) is about 200 μm, and its thermal conductivity could reach about 1200 W/m K. However, there is still no breakthrough in the preparation of millimeter-thick graphite films and PI-derived graphite blocks with large sizes. In the future, as the application range widens, the market demand for high-thermal-conductivity graphite films will be more diversified, and the diverse products will also be developed in the direction of wider thickness and higher thermal conductivity.

The emerging modification treatment and composite technology provide a promising strategy not only to improve the comprehensive performance (e.g., high thermal conductivity and good mechanical flexibility) of PI-derived graphite films but also to prepare a variety of new forms of PI-derived carbon materials with high thermal conductivity. Furthermore, polymer-derived carbon materials with a significant feature of three-dimensional high thermal conductivity can be achieved by novel structure design.

At present, high-thermal-conductivity graphite films have been widely used in smart phones, successfully solving the heat dissipation problem of various electronic products. In the near future, with the development of miniaturization and thinning of electronics, high-thermal-conductivity graphite films and other carbon composites with good flexibility will be promisingly used in the field of thermal management as next-generation heat-dissipation components for highly integrated microelectronics, 5 G wireless communication, and high-power smart devices.

Acknowledgements

We acknowledge support of the publication fee by the National Natural Science Foundation of China (Grant No. 52072275).

Conflict of interest

The authors declare no conflict of interest.

Author details

Guanming Yuan[1,2*] and Zhengwei Cui[1,2]

1 The State Key Laboratory of Refractories and Metallurgy, Wuhan University of Science and Technology, Wuhan, China

2 Hubei Province Key Laboratory of Coal Conversion and New Carbon Materials, Wuhan University of Science and Technology, Wuhan, China

*Address all correspondence to: yuanguanming@wust.edu.cn

IntechOpen

References

[1] Inagaki M. New Carbons Control of Structure and Functions. Oxford: Elsevier Science Ltd; 2000. pp. 35-39. DOI: 10.1016/B978-008043713-2/50010-8

[2] Inagaki M, Ohta N, Hishiyama Y. Aromatic polyimides as carbon precursors. Carbon. 2013;61:1-21. DOI: 10.1016/j.carbon.2013.05.035

[3] Burger A, Fitzer E, Heym M, et al. Polyimides as precursors for artificial carbon. Carbon. 1975;13(3):149-157. DOI: 10.1016/0008-6223(75)90225-0

[4] Inagaki M, Takeichi T, Hishiyama Y, Oberlin A. High quality graphite films produced from aromatic polyimides. In: Thrower PA, Radovic LR, editors. Chemistry and Physics of Carbon. Volume 26. New York: Marcel Dekker, Inc; 1999. pp. 245-333.

[5] Hishiyama Y, Yoshida A, Kaburagi Y. Graphite films prepared from carbonized polyimide films. Carbon. 1992;30(3):333-337. DOI: 10.1016/0008-6223(92)90027-T

[6] Bourgerette C, Oberlin A, Inagaki M. Structural and textural changes from polyimide Upilex to graphite: Part III. Journal of Materials Research. 1993;8(1):121-130. DOI: 10.1557/JMR.1993.0121

[7] Hishiyama Y, Nakamura M, Nagata Y, et al. Graphitization behavior of carbon film prepared from high modulus polyimide film: synthesis of high-quality graphite film. Carbon. 1994;32(4):645-650. DOI: 10.1016/0008-6223(94)90085-X

[8] Kaburagi Y, Hishiyama Y. Highly crystallized graphite films prepared by high-temperature heat treatment from carbonized aromatic polyimide films. Carbon. 1995;33(6):773-777. DOI: 10.1016/0008-6223(95)00009-3

[9] Hishiyama Y, Igarashi K, Kanaoka I, et al. Graphitization behavior of Kapton-derived carbon film related to structure, microtexture and transport properties. Carbon. 1997;35(5):657-668. DOI: 10.1016/S0008-6223(97)00021-3

[10] Zhong DH, Sano H, Kobayashi K, Uchiyama Y. A study of film thickness dependence of the graphitizability of PMDA–ODA polyimide-derived carbon film. Carbon. 2000;38(15):2161-2165. DOI: 10.1016/S0008-6223(00)00075-0

[11] Murakami M, Tatami A, Tachibana M. Fabrication of high quality and large area graphite thin films by pyrolysis and graphitization of polyimides. Carbon. 2019;145:23-30. DOI: 10.1016/j.carbon.2018.12.057

[12] Kato T, Yamada, Y, Nishikawa Y, et al. Carbonization mechanisms of polyimide: Methodology to analyze carbon materials with nitrogen, oxygen, pentagons, and heptagons. Carbon. 2021;178:58-80. DOI: 10.1016/j.carbon.2021.02.090

[13] Murashima K, Kawashima Y, Ozaki S, et al. Modified-edge-support heat treatment method of polyimide for crystalline, large-area, and self-standing ultrathin graphite films. Carbon. 2021;181:348-357. DOI: 10.1016/j.carbon.2021.05.036

[14] Inagaki M, Kang FY. Materials Science and Engineering of Carbon: Fundamentals. 2nd Edition. Beijing: Tsinghua University Press; 2014. p. 265. DOI: 10.1016/C2013-0-13699-9

[15] Bertram A, Beasley K, Torre W. An overview of navy composite developments for thermal management. Naval Engineers Journal. 1992;104(4):276-285. DOI: 10.1111/j.1559-3584.1992.tb01170.x

[16] Zheng Q, Braun PV, Cahill DG. Thermal conductivity of graphite thin

films grown by low temperature chemical vapor deposition on Ni (111). Advanced Materials Interfaces. 2016;3(16):1600234. DOI: 10.1002/admi.201600234

[17] Kato R, Hasegawa M. Fast synthesis of thin graphite film with high-performance thermal and electrical properties grown by plasma CVD using polycrystalline nickel foil at low temperature. Carbon. 2019;141:768-773. DOI: 10.1016/j.carbon.2018.09.074

[18] Hu KS, Chung DDL. Flexible graphite modified by carbon black paste for use as a thermal interface material. Carbon. 2011;49(4):1075-1086. DOI: 10.1016/j.carbon.2010.10.058

[19] Hou SY, He SJ, Zhu TL, et al. Environment-friendly preparation of exfoliated graphite and functional graphite sheets. Journal of Materiomics. 2021;7(1):136-145. DOI: 10.1016/j.jmat.2020.06.009

[20] Thermal protection sheet (Graphite Sheet (PGS)/PGS applied products/ NASBIS). https://industrial.panasonic. com/ww/products/pt/pgs

[21] Wu TS, Xu YL, Wang HY, Sun ZH, Zou LY. Efficient and inexpensive preparation of graphene laminated film with ultrahigh thermal conductivity. Carbon. 2021;171:639-645. DOI: 10.1016/j.carbon.2020.09.039

[22] Chen S, Wang Q, Zhang M, et al. Scalable production of thick graphene films for next generation thermal management applications. Carbon. 2020;167:270-277. DOI: 10.1016/j.carbon.2020.06.030

[23] Song NJ, Chen CM, Lu CX, et al. Thermally reduced graphene oxide films as flexible lateral heat spreaders. Journal of Materials Chemistry A. 2014;2(39): 16563-16568. DOI: 10.1039/C4TA02693D

[24] Li HL, Dai SC, Miao J, et al. Enhanced thermal conductivity of graphene/polyimide hybrid film via a novel "molecular welding" strategy. Carbon. 2018;126:319-327. DOI: 10.1016/j.carbon.2017.10.044

[25] Akbari A, Cunning BV, Joshi SR, et al. Highly ordered and dense thermally conductive graphitic films from a graphene oxide/reduced graphene oxide mixture. Matter. 2020;2(5):1198-1206. DOI: 10.1016/j.matt.2020.02.014

[26] Zhong J, Sun W, Wei Q, et al. Efficient and scalable synthesis of highly aligned and compact two-dimensional nanosheet films with record performances. Nature Communications. 2018;9(1):3484. DOI: 10.1038/s41467-018-05723-2

[27] Liu YJ, Li P, Wang F, et al. Rapid roll-to-roll production of graphene films using intensive Joule heating. Carbon. 2019;155:462-468. DOI: 10.1016/j. carbon.2019.09.021

[28] Yuan GM, Li XK, Dong ZJ, et al. Preparation and characterization of graphite films with high thermal conductivity. Functional Materials. 2015;46(17):17097-17101. DOI: 10.15541/jim20160156

[29] Tuinstra F, Koenig JL. Raman spectrum of graphite. Journal of Chemical Physics. 1970;53(3):1126-1130. DOI: 10.1063/1.1674108

[30] Dutta AK. Electrical conductivity of single crystals of graphite. Physical Review. 1953;90(2):187-192. DOI: 10.1103/PhysRev.90.187

[31] Lavin JG, Boyington DR, Lahijani J, et al. The correlation of thermal conductivity with electrical resistivity in mesophase pitch-based carbon fiber. Carbon. 1993;31(6):1001-1002. DOI: 10.1016/0008-6223(93)90207-Q

[32] Taylor R. The thermal conductivity of pyrolytic graphite. Philosophical Magazine. 1966;13(8):157-166. DOI: 10.1080/14786436608211993

[33] Norley J. The role of natural graphite in electronics cooling. Electronics Cooling Magazine. 2001. http://www.electronics-cooling.com/2001/08/the-role-of-natural-graphite-in-electronics-cooling

[34] Pop E, Varshney V, Roy AK. Thermal properties of graphene: fundamentals and applications. MRS Bulletin. 2012;37(12):1273-1281. DOI: 10.1557/mrs.2012.203

[35] Inagaki M, Kang FY, Toyoda M, Konno H. Advanced Materials Science and Engineering of Carbon. Beijing: Tsinghua University Press; 2013. p. 68. DOI: 10.1016/C2012-0-03601-0

[36] Yuan GM, Cui ZW. Preparation, characterization, and applications of carbonaceous mesophase: a review. In: Ghamsari MS, Carlescu I, editors. Liquid Crystals and Display Technology. IntechOpen; 2020. pp. 101-120. DOI: 10.5772/intechopen.88860

[37] Marsh H. Introduction to Carbon Science. London: Butterworths; 1989. pp. 7-8.

[38] Peng L, Xu Z, Liu Z, et al. Ultrahigh thermal conductive yet superflexible graphene films. Advanced Material. 2017;29 (27):1700589. DOI: 10.1002/adma.201700589

[39] Wang N, Samani MK, Li H, et al. Tailoring the thermal and mechanical properties of graphene film by structural engineering. Small. 2018;14(29):1801346. DOI: 10.1002/smll.201801346

[40] Wang B, Cunning BV, Kim NY, et al. Ultrastiff, strong, and highly thermally conductive crystalline graphitic films with mixed stacking order. Advanced Materials. 2019;31(29):1903039. DOI: 10.1002/adma.201903039

[41] Ma LR, Wang YX, Wang YY, et al. Graphene induced carbonization of polyimide films to prepared flexible carbon films with improving-thermal conductivity. Ceramics International. 2020;46:3332-3338. DOI: 10.1016/j.ceramint.2019.10.042

[42] Wang K, Li MX, Zhang JJ, Lu HB. Polyacrylonitrile coupled graphite oxide film with improved heat dissipation ability. Carbon. 2019;144: 249-258. DOI: 10.1016/j.carbon.2018.12.027

[43] Huang HG, Ming X, Wang YZ, et al. Polyacrylonitrile-derived thermally conductive graphite film via graphene template effect. Carbon. 2021;180:197-203. DOI: 10.1016/j.carbon.2021.04.090

[44] Yan H, Mahanta NK, Wang BJ, et al. Structural evolution in graphitization of nanofibers and mats from electrospun polyimide–mesophase pitch blends. Carbon. 2014;71:303-318. DOI: 10.1016/j.carbon.2014.01.057

[45] Li A, Ma ZK, Song HH, et al. The effect of liquid stabilization on the structures and the conductive properties of polyimide-based graphite fibers. RSC Advance. 2015;5:79565-79571. DOI: 10.1039/C5RA10497A

[46] Tao ZC, Wang HB, Lian PF, et al. "Graphitic bubbles" derived from polyimide film. Carbon. 2017;116: 733-736. DOI: 10.1016/j.carbon.2017.02.044

[47] Ou AP, Huang Z, Qin R, et al. Preparation of thermosetting/thermoplastic polyimide foam with pleated cellular structure via in situ simultaneous orthogonal polymerization. ACS Applied Polymer Materials. 2019;1(9):2430-2440. DOI: 10.1021/acsapm.9b00558

[48] Feng JZ, Wang X, Jiang YG, et al. Study on thermal conductivities of aromatic polyimide aerogels. ACS Applied Materials & Interfaces. 2016;8(20):12992-12996. DOI: 10.1021/acsami.6b02183

[49] Murakami M, Nishiki N, Knakamura K, et al. High-quality and highly oriented graphite block from polycondensation polymer films. Carbon. 1992;30(2):255-262. DOI: 10.1016/0008-6223(92)90088-E

[50] Wang JM, Li QX, Liu D, et al. High temperature thermally conductive nanocomposite textile by "green" electrospinning. Nanoscale. 2018;10:16868-16872. DOI: 10.1039/C8NR05167D

[51] Kausar A. Emerging polyimide and graphene derived nanocomposite foam: research and technical tendencies. Journal of Macromolecular Science, Part A: Pure and Applied Chemistry. 2021;58(10):643-658. DOI: 10.1080/10601325.2021.1934011

[52] Zhang XD, Guo Y, Liu YJ, et al. Ultrathick and highly thermally conductive graphene films by self-fusion. Carbon. 2020;167:249-255. DOI: 10.1016/j.carbon.2020.05.051

[53] Zhu Y, Peng QY, Qin YY, et al. Graphene-carbon composite films as thermal management materials. ACS Applied Nano Materials. 2020;3(9):9076-90887. DOI: 10.1021/acsanm.0c01754

[54] Li YH, Zhu YF, Jiang GP, et al. Boosting the heat dissipation performance of graphene/polyimide flexible carbon film via enhanced through-plane conductivity of 3D hybridized structure. Small. 2020;16(8):1903315. DOI: 10.1002/smll.201903315

[55] Ning Wen, Wang ZH, Liu P, et al. Multifunctional super-aligned carbon nanotube/polyimide composite film heaters and actuators. Carbon. 2018;139:1136-1143. DOI: 10.1016/j.carbon.2018.08.011

[56] Ou XH, Chen SS, Lu XM, Lu QH. Enhancement of thermal conductivity and dimensional stability of polyimide/boron nitride films through mechanochemistry. Composites Communications. 2021,23:100549. DOI: 10.1016/j.coco.2020.100549

[57] Loeblein M, Bolker A, Tsang SH, et al. 3D Graphene-infused polyimide with enhanced electrothermal performance for long-term flexible space applications. Small. 2015;11(48):6425-6434. DOI: 10.1002/smll.201502670

[58] Wang Y, Wang HT, Liu F, et al. Flexible printed circuit board based on graphene/polyimide composites with excellent thermal conductivity and sandwich structure. Composites Part A: Applied Science and Manufacturing. 2020;138(44):106075. DOI: 10.1016/j.compositesa.2020.106075

[59] Luo XH, Guo QG, Li XF, et al. Experimental investigation on a novel phase change material composites coupled with graphite film used for thermal management of lithium-ion batteries. Renewable Energy. 2020;145:2046-2055. DOI: 10.1016/j.renene.2019.07.112

Polyimide: From Radiation-Induced Degradation Stability to Flat, Flexible Devices

Prashant S. Alegaonkar, Vasant N. Bhoraskar
and Sudha V. Bhoraskar

Abstract

Polyimide (PI, PMDA-ODA, $C_{22}H_{11}N_2O_5$, Kapton-H), is a class of polymer, extensively used in microelectronics and space technology, due to its exceptional mechanical, dielectric, and chemical properties. In space, PI heat shield experiences a harsh environment of energetic electrons, ultra-violet radiation, and atomic oxygen, causing degradation and erosion. Radiation-assisted physicochemical surface modulations in PI, in view of understanding and reducing the degradation in laboratory-based systems, are discussed in the chapter. Strategies for the design and development of 2D, flat, and flexible electromechanical devices by swift heavy ion induced bulk modifications in PI are also described. Fabrication of a couple of such devices, including their performance analysis, is presented.

Keywords: polyimide, radiation, degradation, stability, devices

1. Introduction

As one of the high-performance polymers, aromatic polyimides (PIs), find applications in many high-tech fields due to their excellent thermal and oxidative stabilities, high mechanical strength, flexibility, and good dielectric properties. Its importance has been established on the basis of exceptional and versatile properties, which is unparallel to most other classes of macromolecules. There are both thermosetting and thermoplastic PIs. In fact, there are many published comprehensive works on PIs' chemistry, synthesis, characterization, and applications [1–3]. PIs were first synthesized by chemists at Dupont in the 1950s and are popular by the trade name of Kapton. Most often, this aromatic form of PI is used for space applications. Kapton is prepared through condensation polymerization of pyromellitic dianhydride (PMDA) and oxy-di-aniline (ODA).

PIs are used extensively in aerospace, gas separation, memory devices, and in the microelectronics industry. PI also finds application in the electrical industry as insulation coating of electromagnetic wirings. It can withstand the temperature of 425°C for short time exposures without undergoing any degradation or deformation. For this reason, PI finds application in supersonic aircraft and in space vehicles. Most of the materials developed for the space application were initiated during the 1960s, when Sputnik—"Satellite I" was launched by the Soviet Union. Then onwards space programs were developed in various directions. The applications involved the requirements in various earth orbits and interplanetary missions. Earth orbits are classified as low Earth orbit (LEO), between 200 and 800 km, geostationary Earth orbit (GEO) defined at 36,000 km above the equator, and medium Earth orbit (MEO), existing between LEO and GEO. Each orbit is employed for specific applications, like navigation, communication, and for use in Earth observation satellites.

However, prolonged use of spacecraft materials and their exposure to the space environment leads to the degradation of thermal, electrical, mechanical, and optical properties, which can subsequently result in an early mission failure. Materials used in space are exposed to various hazards, which include ultraviolet (UV) radiations and other ionizing radiations (like energetic electrons, protons, and heavy ions) [4]. Ground-simulated research is, therefore, extensively required to safeguard the properties of PI from these radiations [5].

Further, due to the developments in micro and nanoelectronics which are approaching smaller and smaller sizes at increasing production costs, many small- and medium-sized manufacturers have realized the difficulty of survival due to the harsh competition. Here polymeric electronics offers the solution with flexible thin-printed transistor electronics which can provide large-size displays. These nanosized structures are capable of covering the bridge between the "classical" silicon-based, and the new polymer-based electronics. One such development is based on the ion tracks in polymers: similar to those formed in photoresist and SiO_2 (both of which are the vital components of many silicon-based structures). Now silicon/track hybrid structures may be designed inside the polymers. These ion tracks in polymer foils will form a multitude of new interesting applications not only in electronics but also in other fields such as medicine or optics.

Ion tracks [6] are formed in polymers when high-energy ions damage the polymer by the process of chain scission and cross-linking along their passage. As a result of bond breaking in the polymer, gases like hydrogen, CO_2, CO, or CH_4 are liberated along the trajectory. When these gases get emitted out of the surface, they create holes and thus form the tracks which are often called Latent Tracks. After chemical etching and removing the damaged matter using a suitable etchant, these tracks get opened up. They are visible with a microscope and are called Etched Ion Tracks [6]. If such tracks are filled with an appropriate material like silver or gold, they can generate nanowires or nanorods within the polymer [7]. By generating such structures in PI and filling the tracks with silver, Fink et al. [6] have successfully fabricated micro-transformers. Petrov et al. [8] have reported on the effects of time of irradiation on the crystallinity, conductivity, and wall thickness of the nanotubes [8].

The use of PI in electronic applications is also exhibited by its good electrical and dielectric stability, apart from its mechanical strength. In general, polymers are low dielectric constant ($\acute{e}$) materials. Polyimide for example has a dielectric constant equal to ~3.15. Inorganic materials and oxides have relatively higher values of dielectric constant (between 4 and 9). The values of the dielectric constant can be therefore tailored by making composites or by doping an external material into the host polymer. To find suitable applications as an insulator, the material should preferably have a low dielectric constant. In fact, both, the dielectric constant and dielectric relaxation (τ) are especially important in deciding the suitability of the polymer in technological applications. These properties are controlled to some extent by the defects and space-charge, generated, within the folded chains of a polymer. Defects generated during the synthesis also lead to the formation of dipoles within the polymer matrix and they control the dielectric properties. This results in the dielectric-relaxation described by the value of dissipation factor, tan δ, and its relation with τ_0 (the relaxation time for dipole orientation) given by the expression [9, 10]:

$$\tan \delta = 2\,\pi f.\tau_0.S_r / 1 + (2\,\pi f\,\tau_0)^2. \qquad (1)$$

Here, S_r is the relaxation strength which depends on the number density of the dipoles and f is the frequency of the applied electric field.

Due to its high mechanical and thermal stability PI is used as an excellent dielectric material for spacecraft technologies. There are reports [11] whereby using the nano-foam morphology the dielectric constant of PI has been lowered from 3.2 to 2.5 at ambient temperatures, and from 2.9 to 2.3 at 100°C. On the other hand, there are reports [12] where the refractive index (directly related to the dielectric constant) has been increased in fluorinated PI when it is irradiated with 25 keV electrons at a fluence of ~5 × 10^{15} e/cm^2. The dielectric constant is also very influential in controlling the signal transmission and its attenuation when PI is used for packaging in high-speed microelectronic devices [13]. High values of the dielectric constant are useful when the polymers are used as insulating layers in the high voltage capacitors. Such studies have been, therefore, carried out with PI from the point of view of finding its applications in the micro-electronic and optoelectronic fields [14–17]. However, the bulk resistivity of a polymer should be as high as possible to protect its insulating characteristics. These are often the fundamental issues that are aimed at while tailoring the dielectric properties of polymers.

Again, while talking about the space applications of PI its interaction with high energy ionizing radiations becomes quite imperative. Effect of ionizing radiations on polymers, radiation processing of polymers, surface modifications, and use of radiations for nanotechnology [18–22] have been largely reported. In general, the irradiation on polymer leads to the formation of reactive species such as free radicals.

PI is used as a shielding material in different application areas. These include space missions, automotive industries, and nuclear electronics. In these applications, its optical reflectivity also plays an important role. It is also used as a thermally strong insulator in several electronic circuits [23, 24]. There are reports where its thermal and mechanical properties are studied as a function of high energy radiations such as 3 MeV protons [24] and 2 MeV electron irradiation [25]. Effects of 2 MeV electron irradiation have been reported to alter the mechanical properties of PI [26]. The change in the hardness and Youngs Modulus by 4 MeV light ions on different kinds of PIs are also reported [27]. Radiation processing using high-energy electron irradiation is especially preferred for polymers because of their high dose rates and high-energy deposition in lesser time intervals. High processing rates are achieved because a high dose rate also initiates a high rate of cross-linking as oxidation effects are relatively weak at the high dose rates. Moreover, electron irradiation facilities are easily controllable and one can achieve steady dose rates without any interruption of homogeneous energy deposition [28, 29]. In particular, the effects of high-energy pulsed electron irradiation for studying the surface processing of PI need more attention.

In satellites, PI is used for different applications such as shielding material for thermal control, insulating material in electronic circuits, and similar others. The satellites that are designed to operate in the low Earth orbits region (LEO) [30, 31] are exposed to atomic oxygen in addition to other space radiations such as electrons, protons, and UV radiation [32–34]. As the period of exposure to this radiation increases, the physicochemical, mechanical, optical, thermal, and electrical properties of polymers degrade at a faster rate [32]. Effects of Atomic Oxygen (AO) having energies between 5 and 20 eV carries significant importance since the space shuttle in the LEO region is exposed to such atomic oxygen species.

In space, the AO is generated when the short-wavelength ultraviolet radiation (>5.12 eV, <243 nm) dissociates molecular oxygen present in the upper atmosphere [33–35]. The space vehicle normally moves with a velocity of 7–8 km s^{-1} in the LEO region. The energy of the atoms of oxygen is decided by the velocity of the space vehicle. The energy with which the atoms of the oxygen impinge on the spacecraft surface, moving with a high velocity, becomes around 5 eV. In this LEO region, the flux of the atomic oxygen varies from $\sim 10^{12}$ to 10^{15} atoms cm^{-2} s^{-1} [36]. The surface region of the polymer which is used as a shielding material over the space-craft, therefore, receives fluences of $\sim 10^{19}$–10^{22} ions cm^{-2} in a period of about 1 year. During the mission period of the spacecraft in the LEO region, the deteriora-tion observed on the surface of the polymer is therefore mainly correlated to the irradiation effects of the atomic oxygen [37]. The polymer-like PI which is used as a shielding material then gets eroded and loses its weight. The rate of weight loss depends on the flux of the atomic oxygen. In general, the weight loss increases with increasing irradiation period [38–42].

In addition, at Low Earth Orbit, oriented spacecraft experiences asymmetric fluxes because of the interactions between the radiations with the environment. As a result, prior to the exposure to atomic oxygen, the spacecraft comes across other space radiations such as electrons, protons, and UV radiations [43–46]. The shielding material on the spacecraft then interacts with all these radiations during its passage which may be more detrimental for the polymer surface. With increasing time of interaction all the physicochemical and related properties of the polymers degrade at a faster rate [47]. Ionizing radiations produce defects mainly by cross-linkages and chain scission. These processes are assisted by bond breakages or weakening of the bonds resulting in the emission of gaseous species from the surface. Moreover, atomic oxygen is very reactive due to the lone pair of electrons it has. Several interactions can occur over the polymeric surface, such as oxidative reactions with surface atoms or adsorbed molecules, both elastic as well as inelastic scattering recombination, associ-ation, or excitation of species. Atomic oxygen can very strongly react chemically with a surface and cause the formation of volatile oxides from polymers. The surface layer gets eroded. Although PI is a thermally stable polymer it is also seen to get degraded by AO. As a consequence of such degradation, research in this area still remains one of the hot topics with PI. The erosion yield of the polymer is also influenced by other physical parameters like the impact angle, temperature, flux, and total fluence of atomic oxygen, synergistic effects of other radiations, and the energy during the impact of the atomic oxygen [48, 49].

There is no dearth of literature where the effect of one kind of radiation alone has been investigated on the change in the properties of PI. For example, Hill et al. have studied the effects of 3MeV proton radiation on the mechanical properties [24], Plis et al. have investigated the spectroscopic and thermal properties of 2MeV electron irradiation [50], Sasuga et al. report the effect of 2MeV electron radiation on the mechanical properties [26] and Cherkashina et al. have studied the effect of 4MeV light ions on the hardness and Youngs' modulus [51] of various polyamides. In our earlier publication [52] it has been shown that 1 MeV pulsed electron beam can be used to simulate conditions similar to space radiations. The surface and bulk proper-ties of PI were modified to investigate the role of electron fluence. Our group has also reported [53–55] the erosion properties of PI and studied the rate of weight loss as a function of the flux of atomic oxygen generated in Electron Cyclotron Plasma. Recently we have investigated the synergetic effects of electron irradiation on the surface erosion properties of PI by AO. The samples were irradiated with 1 MeV electrons prior to exposing them to the atomic oxygen. The electrons having fluence, ranging from 5×10^{14} to 20×10^{14} electrons/cm^2, were obtained from the electron accelerator whereas, the atomic oxygen was generated in a plasma system. Post-characterization of the irradiated samples, using weight loss measurements, indicated that the extent of erosion by oxygen ions increases if the sample is pre-irradiated with 1 MeV electrons [42]. The paper also reports a larger extent of degradation in the optical transparency when samples were pre-irradiated by electrons.

The chapter highlights the effects of radiations (electrons and gamma radiations) in multiple ways. It is used in modifying the dielectric properties and also for elemental doping by radiation-enhanced diffusion process. The surface erosion properties by atomic oxygen, obtained by a ground-based electron cyclotron resonance plasma simulator are also discussed. The use of swift heavy ions in fabricating the micro-electronic device is also presented. The results of different experiments carried out with PI are sequentially described in separate sections.

2. Effect of 1-MeV electron irradiation on dielectric function: boron and fluorine diffusion

Studies related to the dielectric measurements have been extensively carried out in our group and are reported in our earlier communication [53] (Alegaonkar APL 2002). Here, we have chosen a polyimide ($C_{22}H_{10}N_2O_5$, PMDA-ODA, Kapton-H) sheet having a thickness of ~50 mm. The electron irradiation was carried out with a microtron accelerator which was used in the pulsed mode of operation by varying the electron fluence over the range of 10^{14}–10^{15} e/cm^2. The weight loss of the PI samples after electron irradiation, having dimensions of 12 mm × 12 mm, was monitored. All the electron-irradiated samples were subjected to dielectric relaxation measurements over a frequency range of 100 Hz–7 MHz. The values of dielectric constant $\acute{\epsilon}$, and the dielectric loss $\grave{\epsilon}$, were also estimated at each frequency by using the measured values of the capacitance and tan δ. Apart from this the value of the refractive index has been monitored at a wavelength of 622.8 nm, by using an ellipsometer.

The paper also discusses the effect of boron diffusion on the dielectric properties of the polyimide films and compares the effects with those of electron irradiated films. Gamma irradiation from Co-60 with a dose rate in the range from 6 to 48 Mrad was used in the process of radiation-assisted diffusion using BF$_3$ solution. The depth distribution of boron inside the film was studied using the technique of neutron depth profiling [56] and the presence of fluorine was confirmed by X-ray photoelectron spectroscopic measurements. Moreover, it was revealed that boron and fluorine had diffused from both the surfaces up to a depth of ~2 μm on account of gamma irradiation.

The paper reported that the frequency distribution of dielectric loss, $\grave{\epsilon}$, and the dielectric constant, $\acute{\epsilon}$, were influenced by the fluence of irradiation Φ. The measured refractive index has also been shown to depend on the electron fluence. The graphical representation of these parameters is shown in **Figure 1** (reproduced with permission Alegaonkar et al. [53]). The paper also highlights a small shift in the second peak in **Figure 1(a)** towards lower frequency and correlates it to an increase in the relaxation period with increasing electron fluence. This is governed by the relation: $\omega \cdot \tau_0 \cdot T \sim 1$ [1, 57]. The loss factor actually depends on the conductivity of the material. The electrical conductivity in PI, as in most of the polymers, is governed mostly by the mechanism of hopping of charge carriers from one defective site into another. Since electron irradiation is associated with the generation of defect sites it is influenced by the fluence of irradiation. Effectively the number of electrons that can undergo transition has increased with increasing electron fluence. As a result, the dielectric loss, $\grave{\epsilon}$, which is associated with electron transition, also

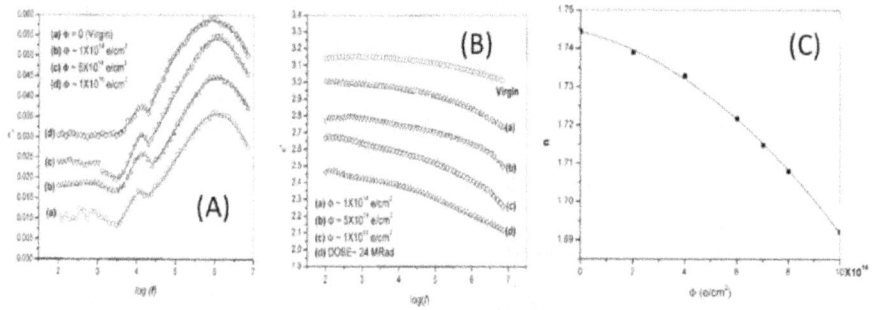

Figure 1.
(A) Variations in dielectric loss, ĕ, with log(f) for PI samples irradiated with 1 MeV electrons at different fluences, Φ. (B) Variations in dielectric constant, ĕ, with log (f) for PI samples irradiated with 1 MeV electrons at different fluences, Φ [plots (a)–(c)] and for a PI sample immersed in BF₃ solution and irradiated with Co-60 gamma rays [plot (d)]. (C) For 1 MeV electron-irradiated PI samples, variation in refractive index, n, with electron fluence, Φ. Reproduced with permission from Alegaonkar et al. [53].

increases with increasing electron fluence. Here, the dielectric constant, ($\varepsilon = 3.15$) of virgin PI is reported to be independent of frequency, whereas for the electron irradiated PI it is seen to decrease as the frequency of the applied electric field increases. The lowest value of the dielectric constant is reported to be 2.4 (at $\sim$7 MHz). For the boron-and-fluorine doped PI, the lowest value is seen to be 2.1 (at $\sim$7 MHz). The refractive index, ($\eta = 1.7444$) of virgin PI has also been reported to decrease with electron fluence reaching the lowest value of $\sim$1.6921, at an electron fluence of 1×10^{15} e/cm². The lowering of the dielectric constant has been correlated to the formation of π-electron clouds resulting from chain scission and cross-linking of molecular chains [2, 6, 7]. The polarization is aligned in the direction of the molecular chains. As a consequence, the refractive index in a direction perpendicular to the plane of the sample is also affected.

The paper also reports a measurable weight loss arising from the emission of gases such as oxygen, hydrogen, etc., from the PI matrix during irradiation. Thus, in this case, the density of the PI has been shown to decrease with increasing electron fluence. The result has been reproduced in **Figure 2**. It shows that the percentage of weight loss, W_L, increases with increasing electron fluence.

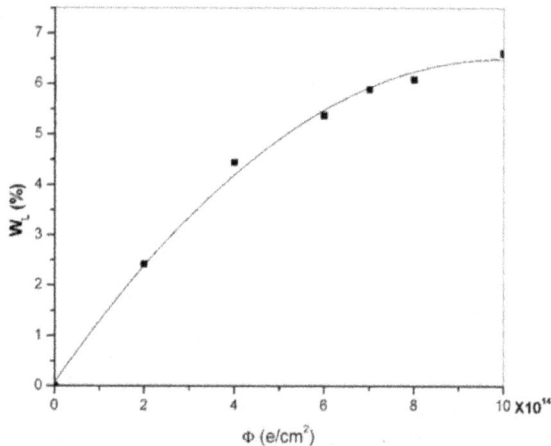

Figure 2.
For 1 MeV electron-irradiated PI samples, variation in weight loss, W_L (%), with electron fluence, Φ. Reproduced with permission. (Reproduced from Alegaonkar et al. [53]) ©Copy right 2002, Applied Physics Letters.

Apart from the irradiation with electrons, our group has also studied [58] the influence of doping the PI with boron and fluorine as stated in the earlier paragraph. Even though boron and fluorine were diffused up to a depth of ~2µm from each side, they lowered the dielectric constant of ~50 µm thick PI quite effectively. The variations in the dielectric constant, $\acute{e}$, and dielectric loss $\breve{e}$, with frequency, were almost similar to those irradiated with 1 MeV electrons at a fluence of 10^{15} e/cm². These results are important in view of finding the application of 50-µm thick PI, where lower values of dielectric constant are preferred.

In another report from our group [59] phosphorus, boron, and fluorine were doped into the PI films by the radiation-assisted diffusion. The dielectric constant, $\acute{e}$, of the PI, was again tailored over a small range around the value of 3.15.

3. Dielectric parameters of phosphorus and fluorine diffused PIs

Similar to the work which has been discussed in the previous section relating to the diffusion of boron into the surface layer of PI, we have also carried out investigations by doping phosphorous and compared the results with those obtained by doping with boron. The results have been reported in our earlier communication [59]. 50 µm thick PI samples of 15 mm × 15 mm were doped with phosphorous and fluorine by using the technique of radiation-assisted diffusion process. Solutions of H_3PO_4 and BF_3 were used to diffuse the elements of phosphorous and B/F into the matrix of PI. The doped samples were designated as S_P, in the case of P doping and $S_{B/F}$ in the case of boron and fluorine doping. Similarly, the unirradiated virgin PI samples are designated as S_V. Different doses of gamma rays, ranging from 64 kGy to 384 kGy from Co-60 were used as the source of radiations.

X-ray photoelectron spectroscopy (XPS) was used to confirm the presence of phosphorus in S_P and fluorine in $S_{B/F}$. The atomic concentrations (%) found from these measurements were as shown in **Table 1**. The paper describes the experimental results of the Rutherford Backscattering Technique which provided the depth of the diffusion of the phosphorus and fluorine in the PI samples. The RBS spectra for the S_P and the $S_{B/F}$ samples are as shown in **Figure 3(a)** and **(b)**, respectively. The RBS spectrum was deconvoluted and the elements present in the PI surface were identified. The average depth of diffusion of phosphorus or fluorine atoms in the PI, estimated at ~10% concentration relative to the peak concentration at the surface, was ~3 µm.

The reported RBS spectra have indicated the presence of phosphorus and fluorine as has been reproduced in **Figure 3**. The area under the RBS peaks has been shown to be proportional to the number of phosphorus or fluorine atoms diffused in

Dose of gamma-rays dose (kGy)	Atomic percentage (%)	
	Phosphorus	Fluorine
64	0.7	1.05
96	1.12	1.68
128	1.5	2.22
192	1.82	2.66
288	2.03	2.92
384	2.21	3.12

Table 1.
The atomic percentage of phosphorus in S_P and fluorine in $S_{B/F}$ samples for different doses of Co-60 gamma-rays, as estimated from the EDAX measurements.

Dose of gamma-rays dose (kGy)	Atomic percentage (%)	
	Phosphorus	Fluorine
64	0.7	1.05
96	1.12	1.68
128	1.5	2.22
192	1.82	2.66
288	2.03	2.92
384	2.21	3.12

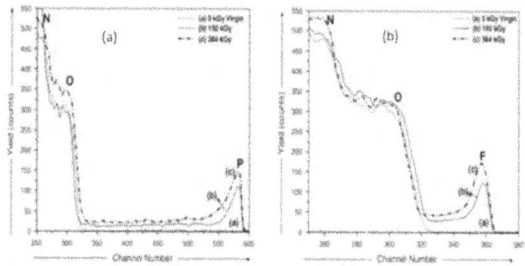

Figure 3.
(a) and (b) RBS spectra of the PI samples which were immersed in H_3PO_4 solution and BF_3 solution respectively. The samples were irradiated with Co-60 gamma-rays at a dose (a) 0 kGy; virgin, (b) 192 kGy and (c) 384 kGy. (Reproduced from [58]) Reproduced with permission ©Copy right 2006, Journal of Physics D.

the PI sample. In addition, the concentration of phosphorus or fluorine (relative %) is shown to increase with an increasing dose of Co-60 gamma-rays. From the results of the EDAX measurements shown in **Table 1**, Riyadh et al. [58] have reported that the atomic percentage of phosphorus, as well as that of fluorine, has increased with the dose of Co-60 gamma-rays. The results have been discussed in view of the defects induced in PI as an effect of gamma irradiation, which includes vacancies, cross-linking and bending of bonds, and formation of voids due to the diffusion of gaseous molecules [13]. The gases such as H, O, and C are normally evolved as an effect of radiolysis. The process of diffusion of external atoms into polymer surface is normally explained using the model based on free volume [58, 59].

Here, the depth of diffusion of the phosphorus and fluorine atoms is reported to be ~3μm. The authors [58] have further analyzed the results with the XPS spectrum (not shown here). The analysis is reported to have further indicated that phosphorus and boron atoms have formed bonds with oxygen atoms [60] in PI, whereas fluorine atoms have formed bonds with the carbon atoms [61] in the surface region. Here, the concentration, Cs, of the phosphorus and fluorine (Cs in relative %) in the S_P and $S_{B/F}$ samples have been obtained after normalizing the intensity of each XPS peak with respect to the corresponding photoelectric cross-section. The variations in the phosphorus and fluorine concentrations with the dose of Co-60 gamma-rays were determined which are represented in **Figure 4(a)** [62].

The dielectric constant, $\acute{\epsilon}$, estimated over the frequency range 100 Hz–7 MHz. are shown in **Figure 4(b)**. The spatial distribution of boron was studied with the

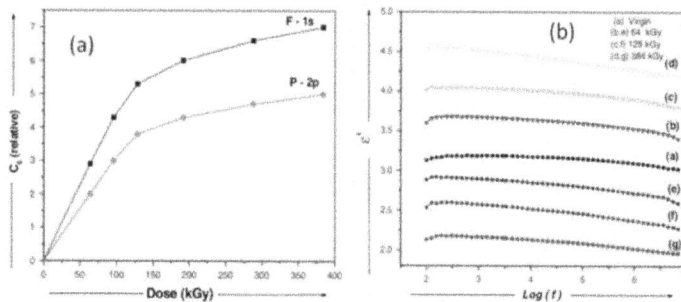

Figure 4.
Variations in the relative concentrations of phosphorus and fluorine with the dose of Co-60 gamma-rays. Variations in the dielectric constant, $\acute{\epsilon}$, with frequency, f, for the PI samples doped with phosphorus, boron, and fluorine at different doses of Co-60 gamma-rays over the range ~64–384 kGy; (i) virgin (a), (ii) phosphorus-doped (b), (c) and (d) and (iii) fluorine and boron-doped (e), (f) and (g). Reproduced with permission [2006] ©Copy right 2002, Nuclear Instruments and Methods–B. (Reproduced from [58]).

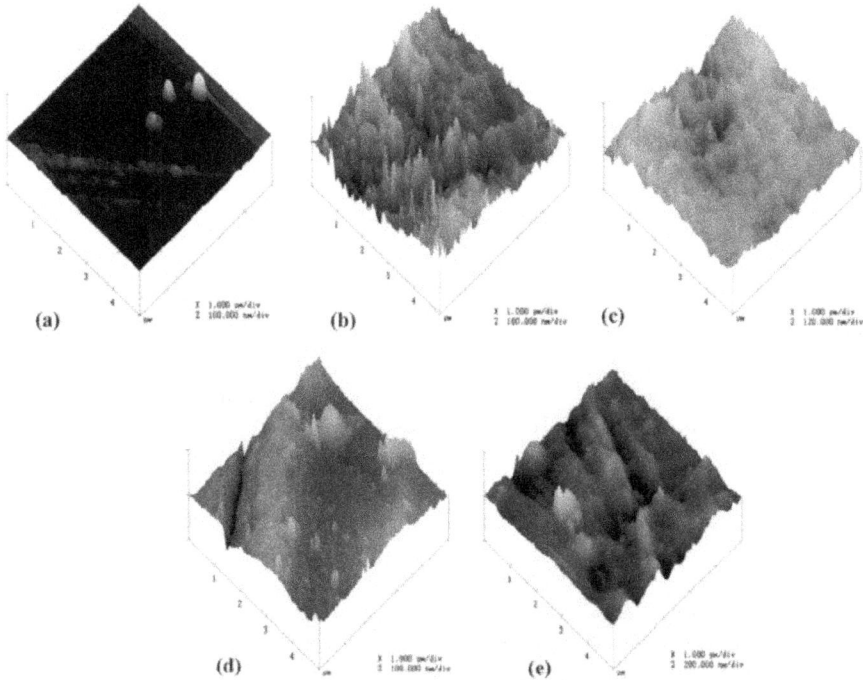

Figure 5.
AFM photographs showing surface images of the PI samples (i) (a) virgin, (ii) doped with phosphorus at gamma-ray dose of (b) 64 kGy, (c) 384 kGy and (iii) doped with boron and fluorine at gamma-ray dose of (d) 64 kGy, (e) 384 kGy. Reproduced with permission [59] © Copyright 2006, J. of Phys. D. [Reproduced from [59]].

neutron depth profiling technique using the nuclear reaction for ^{11}B (n, α) ^{7}Li. The experimental method was similar to that reported earlier for HDPE [57, 58].

The paper (reproduced from [58]) also reports the surface morphology using the atomic force microscope technique, in the non-contact mode. A few typical recorded surface images are reproduced in **Figure 5**. The micrograph (a) is for virgin S_V, and the micrographs (b) and (c) are for the S_P samples and (d) and (e) are for the $S_{B/F}$ samples.

The AFM images reproduced in **Figure 5**, are indicative of the change in surface morphology of PI as a result of doping with phosphorus, boron, and fluorine The RMS values of the surface roughness as reported for the S_V, S_P, and $S_{B/F}$ samples were 8.1 nm, 10.4 nm, and 16.7 nm respectively. As compared with the virgin, the surface roughness of the PI is reported to increase marginally after phosphorus doping but significantly after boron and fluorine doping. The authors have found that the mechanical properties of phosphorus, boron, and fluorine diffused PI were almost similar to those of the virgin PI.

The paper also reports the discussion on the variation of the dielectric constant $\acute{\epsilon}$, the density, ρ, and the polarizability, α, which are related [63] through the relation:

$$\left(\frac{\acute{\epsilon} - 1}{\acute{\epsilon} + 2}\right) \frac{M_w}{\rho} = \frac{N_A \alpha}{3\epsilon_0} \qquad (2)$$

Where, ϵ_0 is the permittivity of free space, M_w is the molecular weight and, N_A is Avogadro's number and ρ is the density. In general, the dielectric constant, $\acute{\epsilon}$ increases by increasing the degree of polarizability, α, which depends on density, ρ, cross-linking, free volume, and chemical coupling. Here, the paper discusses the effect of the dopant such as P, which, being a donor can get latched in between the conjugate chains of PI. Radiation-induced intra-chain coupling of phosphorus atoms thus effectively increases the dielectric constant, $\acute{\varepsilon}$ of the PI.

It is further assumed that as a result of fluorine doping, the free volume of the PI must have increased because an appreciable number of the hydrogen atoms are replaced by fluorine atoms which are larger in size. It is known that the electric polarizability of a C—F bond is lower than that of the C—H bond [14]. The electronic polarization is assumed to have decreased with an increasing number of fluorine atoms in the PI. The plots in **Figure 4(b)** is reproduced from [58] indicates that the dielectric constant, $\acute{\varepsilon}$, decreases with the increasing concentration of fluorine, but it is seen to increase with the increasing concentration of phosphorus in the PI. Boron is also assumed to increase the free volume in the chains of PI. As a result, the number of polarizable groups per unit volume has decreased with the increasing concentration of fluorine or boron atoms in PI.

4. Bulk and surface modifications by 6 MeV electron radiation

For 6 MeV irradiation, Mathakari et al. [64] used a PI sheet with a density ~ 1.43 gm/cm^3 and thickness of 50 μm. The sheet was cut into pieces of size 100 mm × 15 mm × 50 μm for stress-strain studies. Whereas for other characterizations, thin films of size 15 mm × 15 mm × 50 μm were used. The samples were irradiated with a 6 MeV electron beam obtained from the Race track microtron laboratory of the Department of Physics, University of Pune, India. The irradiation was performed in the air under normal thermodynamic conditions. The beam diameter was kept ~ 15 mm and the strips were irradiated at the central location, for mechanical measurements. Whereas, for other samples of dimension 15 mm × 15 mm × 50 μm, they were exposed to the uniform electron beam. Electron fluence was varied in the range of 1–4×10^{15} electrons/cm^2. Stopping power ~ 1.836 MeV cm^2/g and range ~ 2.36 cm for 6 MeV electrons in PI was calculated from the ESTAR program. Calculated doses from the stopping power and range were estimated to be 294–1176 kGy. Following this, pre and post-irradiated samples were characterized by various techniques. The irradiated strips used subjected to stress-strain measurements, carried out using a tensile testing machine (UTM). Testing parameters such as load cell, strain rate, and gauge length were, respectively, kept 100 kg, 20 mm/min, and 50 mm. For each pristine and irradiated, an average of measurements on four samples was taken. From obtained stress-strain profiles, numerous mechanical parameters like tensile strength, percentage elongation, strain energy, and Young's modulus were calculated. By measuring coordinates of breaking points on stress-strain profiles, the tensile strength, and percentage elongation were calculated, whereas, strain energy was calculated from the measured stress-strain curve area. Young's modulus was computed from the initial slope of stress-strain curve.

4.1 Mechanical analysis

Figure 6 displays recorded stress-strain profiles for PI irradiated with 6 MeV electrons. The stress-strain curve was distinctly becoming large with an increase in electron fluence and indicative of enhancement in mechanical properties of PI by electron irradiation.

Calculated mechanical parameters like tensile strength, % elongation, strain energy, and Youngs' modulus (from stress-strain profiles) are, respectively, shown in **Figure 7(a–d)**. Notably, tensile strength, percentage elongation, and strain energy is

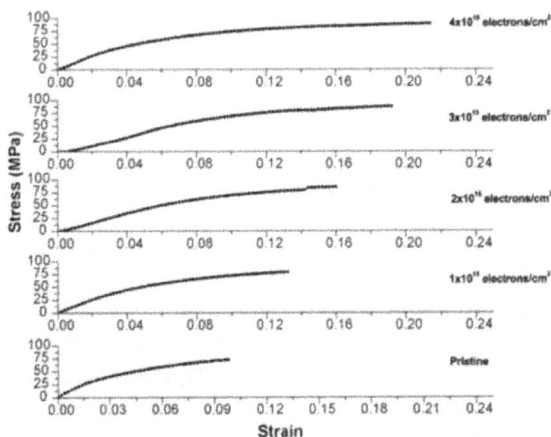

Figure 6.
Stress-strain profiles for pristine and 6 MeV electron irradiated PI. Reproduced with permission [Mathakari et al. [64] ©Copy right 2010, Materials Science and Engineering–B].

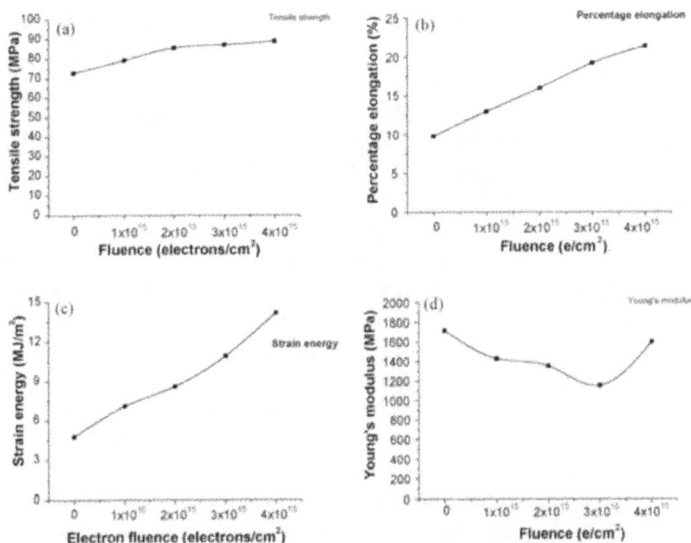

Figure 7.
Recorded profiles of (a) tensile strength, (b) percentage elongation, (c) strain energy, and (d) Young's modulus as a function of fluence for irradiated PI. Reproduced with permission Mathakari et al. [64] ©Copy right 2010, Materials Science and Engineering–B.

observed to be increased from the pristine values of 73–89 MPa, 10–22%, and 4.75–14.2 MJ/m^3, respectively, @ maximum value of fluence (4 × 10^{15} electrons/cm^2) with, corresponding, relative % increments noted to be 22%, 120%, and 199%. Enhancement in mechanical properties is attributed, fundamentally, to change in cross-linking. Upon irradiation, both scission and cross-linking take place in polymer simultaneously, and depending upon polymer, radiation type, and amount of dose rate scission and cross-linking, one or the other may dominate. From the obtained results, Mathakari et al. [64] were of the opinion that in their case cross-linking had dominated over scissioning. The dose rate of 6 MeV electrons was additionally high ~2000 kGy/h. In such a case scenario radiation density induced large area spurs (zones enriched with highly reactive free radicals) overlapping each other. This results in free radicals cross-linking. Further, depending upon dose rate and extent of radiation-induced oxidation; the dominance of scission or cross-linking takes place. But the oxidative changes are weak when dose rates are high [21]. The presence of a large concentration of carbonyl groups in the PI monomer may offer high resistance to oxidation [65]. Thus, the high dose rate and carbonyl-assisted oxidation resistance of PI resulted in extremely low oxidative changes. And hence, crosslinking has dominated over scissioning. Notably, Mathakari et al. [64] observed enhancement in tensile strength, % elongation, strain energy, whereas, Youngs' modulus was found to be decreased, marginally, from the pristine value of 1700–1160 MPa @ 3×10^{15} electrons/cm^2 and then increased up to 1600 MPa @ 4×10^{15} electrons/cm^2. The overall decrease was reported to be 6% and attributed to an increase in % percentage elongation which was considerably larger than the increase in the value of tensile strength. This lead to a conclusion that 6 MeV electron irradiation has considerably enhanced chain dynamics in PI.

4.2 FTIR and UV-vis spectroscopy

Figure 8(a) displayed FTIR spectra recorded for pristine and electron irradiated PI@ different fluences. Peak @ 3085 cm^{-1} is assigned to C H stretching, carbonyl (CO, stretching) in PMDA is located between 172 and 1781 cm^{-1}, hydroxyl (OH) @ 3630 cm^{-1} whereas 1260–1289 cm^{-1} showed C—O—C str. in monomer ODA. Peaks @ 3486 and 1602 cm^{-1} are assigned to N—H str. and bend, respectively. The cyclic imide (C—N—CO) group was scrutinized using peaks that appeared @ 1376–1390 (C—N stretch). The imide stretching was observed @ 1380 and 1117 cm^{-1} and corresponding deformation @ 725 cm^{-1} [66, 67]. Mathakari et al. noted that the absorbance of almost all the bonds remains intact upto $\sim 4 \times 10^{15}$ electrons/cm^2. That revealed scission and crosslinked products have the same absorption spectra as that of pristine PI. More importantly, absorbance at 1725–1781 cm^{-1} corresponding to carbonyl groups remained unchanged till maximum fluence which is indicative of the fact that PI is not oxidized after irradiation.

Further to reconfirm the absence of oxidation effects; **Figure 8(b)** shows only marginal variations in oxygen contents.

Figure 9 shows UV-vis spectra for pristine and electron irradiated PI. They noted, no marked changes in absorption edge and overall absorbance between 700 and 200 nm regions. By and large, PI showed a gradual red shift in the absorbance edge and overall increase in the absorbance due to carbonization and oxidation that confirmed bond structure retention of PI that has not undergone oxidation after irradiation.

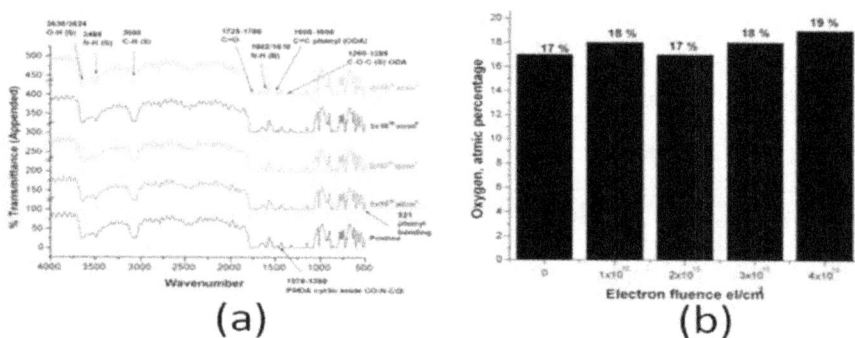

(a)   (b)

Figure 8.
Recorded (a) FTIR spectra and (b) for pristine and 6 MeV electron irradiated PI. Reproduced with permission Mathakari et al. [64] ©Copy right 2010, Materials Science and Engineering–B.

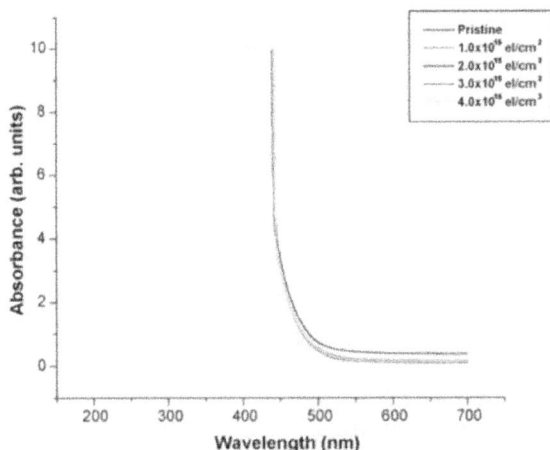

Figure 9.
Recorded UV-vis spectra for pristine and 6 MeV electron irradiated PI @ various fluences. Reproduced with permission Mathakari et al. [64] ©Copy right 2010, Materials Science and Engineering–B.

4.3 Contact angle studies

In **Figure 10**, reported by Mathakari et al., photographic images of distilling water droplet of volume ~10 µl onto the surface of pristine and electron irradiated PI (@4 × 10^15 electrons/cm² is shown. Data of contact angles, roughness, adhesion work, and area fraction of liquid-solid interface @ pristine and electron irradiated PI is provided in **Table 2**. They noted that contact angle decreased from its original (pristine) value of 59–32° @4 × 10^15 electrons/cm². The observed reduction is ~46% and indicative of significant surface modifications in PI surface by 6 MeV electrons. The work of adhesion, roughness and area fraction of the liquid-solid interface is found to be increased with the fluence.

From data of contact angle measurements, a significant reduction in contact angle is observed and, correspondingly, increase in work of adhesion and fractional area @ liquid-solid interface has, mainly, attributed to the radiation-induced surface roughening. By Wenzel formulation: $\cos \Theta' = r \cos \Theta$, for hydrophilic surfaces contact angle decreases with an increase in surface roughness. The observed increase in roughness reveals that there is a greater amount of fractional are available for the contact @ solid/liquid interface [68, 69]. Both the enhanced parameters reduced the contact angle. Further, irradiation of PI surface may also cause the

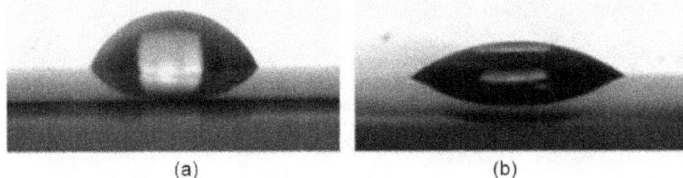

(a) (b)

Figure 10.
Photographic images of contact angle measurements displaying distilled water droplet of 10 µl volume onto the surfaces of (a) pristine and (b) 6 MeV electron irradiated PI@ 4 × 10^15 electrons/cm². Reproduced with permission Mathakari et al. [28] ©Copy right 2010, Nuclear Instruments and Methods–B.

Sr. no.	Fluence ($\times 10^{15}$)	Contact angle (°)	Roughness (r) cos $\Theta' = r \cos \Theta$	Adhesion work ($W_A = \gamma_{Lg}$ (1 + cos Θ))	Area fraction (f) @ liquid-solid interface cos $\Theta' = f \cos \Theta + f - 1$
1	0.0	59	1.00	110.3	1.000
2	1.0	51	1.22	118.6	1.075
3	2.0	49	1.27	120.6	1.093
4	3.0	35	1.59	132.4	1.201
5	4.0	32	1.65	134.5	1.210

γ_{Lg} = surface tension of liquid-gas interface = 72.8 mJ/m². Θ = contact angle of the flat pristine sample. Θ' = contact angle of the rough sample after irradiation.
Reproduced with permission Mathakari et al. [28] ©Copy right 2010, Nuclear Instruments and Methods–B.

Table 2.
Data of contact angles, work of adhesion, and area fraction @ liquid-solid interface of pristine and 6 MeV electron irradiated PI.

opening of imide rings that lead to carboxylic acid and carboxamide formation. They might have played a significant role in reducing contact angle. Thus, a decrease in contact angle is indicative of the hydrophilic nature of PI surface facilitated by irradiation. Such peculiar surface properties with increased hydrophilicity can be used for a number of applications such as grafting, adsorption, or adhesion of organic or inorganic species onto the surface.

4.4 AFM and profilometry investigations

Thus, analysis of Mathakari et al., showed electron irradiation-induced surface roughening is responsible to reduce contact angle. This is reconfirmed in AFM and profilometry. **Figure 11** (upper pan) displays AFM images and (lower part) surface profilograms of pristine and electron irradiated PI@ 4×10^{15} electrons/cm². They

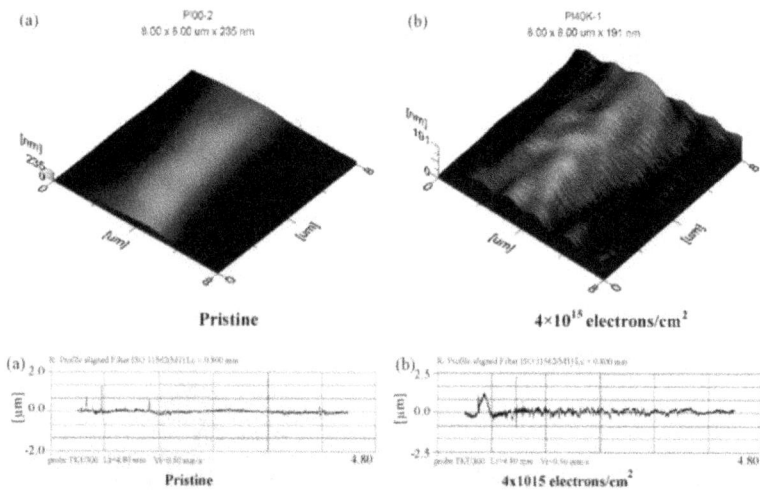

Figure 11.
Upperpart: AFM images (a) pristine and (b) 6 MeV electron irradiated PI at the fluence of 4×10^{15} electron/cm². Lower portion surface profile grams of (a) pristine and (b) 6 MeV electron irradiated PI at the fluence of 4×10^{15} electrons/cm². Reproduced with permission Mathakari et al. [64] ©Copy right 2010, Materials Science and Engineering–B.

noted the surface roughness enhancement after electron irradiation. And the average magnitude of roughness (Ra) is calculated from the profilograms as shown in **Figure 11**. The roughness is found to increase from 0.06 to 0.1 μm and is attributed to the electron beam-induced evolution of gases from the PI surface. The evolution of gases takes place in different regions to a different extent. Chain dynamics seems to be prominent on the surface causing scission and crosslinking in PI modifying surface properties. At bulk, this has implications on change in free volume and induces void type defects, which diffuse up onto the surface enhancing roughness [70].

In the next section, irradiation effects of oxygen ions on PI are discussed and vis-a-vis compared with fluorinated ethylene propylene (FEP).

5. Plasma effects: atomic oxygen

5.1 ECR plasma system

In a study carried out by Riyadh et al., atomic oxygen plasma ions are produced in Electron Cyclotron Resonance (ECR) plasma system. The system is comprised of two stainless steel chambers, one for generating microwave power-induced gas plasma and the other for sample irradiation by plasma. The plasma chamber is configured at a height of ~150 mm and dia. ~125 mm and also function as a microwave cavity resonator in TE111 mode. A magnetron (power ~ 500 W, freq. 2.45 GHz) was coupled to a plasma chamber through waveguides. The magnetic field required for generating the ECR plasma was produced via pair of solenoid coils, mounted around the plasma chamber. The irradiation chamber at a height of ~300 mm and a diameter of ~200 mm has numerous ports that were used for mounting different systems such as reaction gas injector, mechanical support to the samples holder, feed through for electrical measurements, vacuum gauges, window view, etc. The coupling between irradiation and plasma chamber was done in such a fashion that the ions produced in ECR plasma could directly interact with the samples mounted onto the irradiation chamber. In order to measure the energy and flux of oxygen ions, a three-grid retarding field analyzer (RFA) was mounted on the chamber. In both chambers, a pressure of $\sim10^{-6}$ mbar was maintained with an appropriate vacuum system. A schematic design and irradiation setup for the ECR plasma system is shown in **Figure 12**.

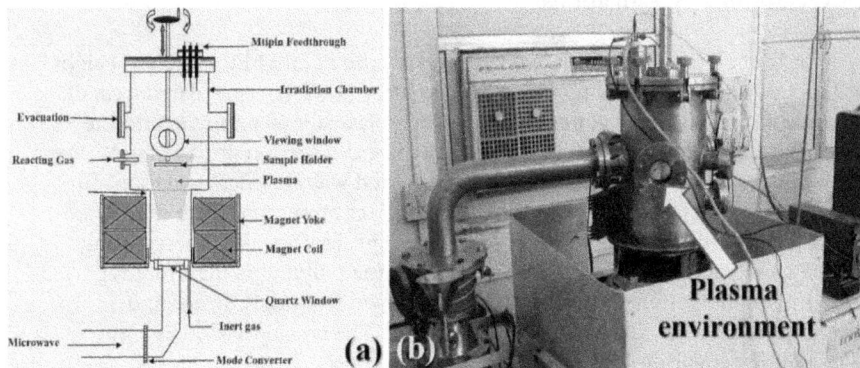

Figure 12.
(a) A schematic design, and (b) experimental set-up of ECR Plasma system. Reproduced with permission Riyadh et al. [55] ©Copy right 2006, Radiation Effects and Defects in Solids.

5.2 Sample preparation

For plasma treatment, PI and FEP samples were obtained by cutting ~50 μm thick sheets into small pieces of dimension ~15 × 15 mm². After cleaning and drying, the samples were stored in a dry and clean atmosphere. Before plasma treatment, the weight of each (PI and FEP) sample was measured by Riyadh et al. using a digital balance, with an accuracy of ±1 μg.

5.3 Irradiation with oxygen ions

After obtaining base pressure ~10^{-6} mbar and making electrical connections, pure oxygen gas (99.9%) was inflected into the plasma chamber via a mass flow controller. Numerous experimental parameters like oxygen flow rate, microwave power, axial magnetic field, oxygen pressure, etc., were optimized to obtain a stable oxygen plasma into the reactor chamber. Energy and oxygen flux of plasma ions at certain positions along with the axial chamber position and @different pressures were measured using three grid RFA. Measurements were repeated and the system was calibrated for ion energy and flux at different positions of plasma. ECR plasma was also characterized by quadrupole mass spectrometry. The intensity of atomic oxygen ions was found to be almost one order of magnitude higher than that for molecular oxygen ions, at all the energies of oxygen ions. Since the abundance of atomic oxygen ions was very high compared to molecular oxygen ions. Four samples of PI and FEP were mounted on the sample holder such that the separation between two conjugate samples was ~3 mm. Samples were, subsequently, irradiated with plasma at 8 × 10^{-3} mbar, at a position where the energy of plasma species was ~10^{-14} eV. However, ion flux was maintained at maxima with an energy of ~12 eV. The flux of ions with ~12 eV of average energy was ~5 × 10^{13} ions cm^{-2} s^{-1}. The sample holder was mounted onto insulating support that was functioned as a Faraday cage. A current integrator, connected to the sample holder, was used to measure a current due to ions received by samples during ion irradiation. Four samples at a time were exposed to oxygen ions at a flux of ~5 × 10^{13} ions cm^{-2} s^{-1}. Ion fluence was varied by changing the time of irradiation that was estimated from the charge received by the samples during ion irradiation. Ion fluence was varied from sample to sample in a range ~5 × 10^{16} to 2 × 10^{17} ions cm^{-2} with time. Plasma-treated samples were characterized by various techniques.

5.4 Weight-loss investigations

Riyadh et al. reported in **Figure 13** that, for both PI and FEP samples, weight loss was increased with increasing plasma exposure, t (s). They speculated a small amount of surface atoms got eroded out when plasma ions impinge onto the polymer surface. However, they noted that the cross-section of this process was small at the set ion energy. Another process noted was chemical reactions. In this atomic oxygen, ions interact with near-surface atoms and slash the weak bonds. This results in the evolution of gaseous species from polymers during plasma treatment. The observed weight loss in the polymer is therefore reported to be a combined effect of atoms that are removed from the surface and redeposited onto the surface during irradiation. The erosion yield (Ey) was calculated by equation [2]:

$$E_y = \frac{\Delta M}{(\rho A \Phi t)} \tag{3}$$

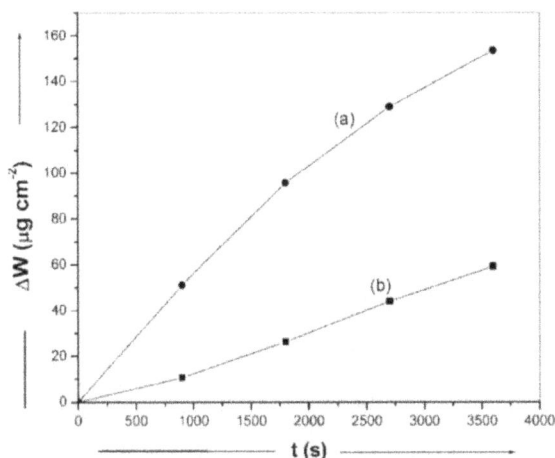

Figure 13.
Weight loss (ΔW) (μg/cm²) variations with time of exposure t (s) to the atomic oxygen ions at a flux of ~5 × 10¹³ ions cm⁻² s⁻¹ (a) PI (b) FEP. Reproduced with permission Riyadh et al. [55] ©Copy right 2006, Radiation Effects and Defects in Solids.

where, Φ is incident ion flux, M is mass loss (gm), ρ is matrix density (1.42 gm cm⁻³(PI) and 2.15 gm cm⁻³(FPE)), A is sample area (2.25 cm²), and t is the exposure time(s). In FEP weight loss is reported due to rupturing of CF_2 bonds and defluorination. After fluorine ejection, new bonds like C—O and C=O are possible to form in the surface region. Moreover, oxidation of C—F bonds is endothermic [33] for these processes. The erosion yield in FEP may be governed by various radiation-induced processes like breaking of chemical bonds and new bond formation, adsorption, or bonding of oxygen atoms to the polymer matrix. In this, the reaction efficiency of atomic oxygen with polymer surface depends upon the chemical composition and bond strength. The average erosion yield (Ey) for the PI is estimated to be ~7 × 10⁻²² cm³ per atom and ~1.35 × 10⁻²² cm³ per atom for FPE. Their results indicated that erosion yield for FEP is ~5 times lower than that for PI.

5.5 Morphological analysis: SEM

In **Figure 14** micrographs reported by Riyadh et al. shows image (1a) and (2a), revealing surface regions of virgin counterparts of PI and FEP. They were reported to be smooth and featureless. However, after plasma treatment, the surface texture

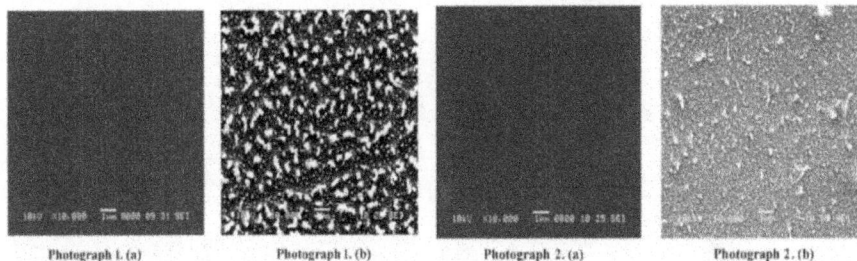

| Photograph 1. (a) | Photograph 1. (b) | Photograph 2. (a) | Photograph 2. (b) |

Figure 14.
Micrograph 1 is recorded SEM for (a) virgin PI, and (b) plasma-treated PI. Image 2: for (a) virgin FEP and (b) plasma-treated FPE. (@Ion fluence ~2 × 10¹⁷ ions cm⁻²). Reproduced with permission Riyadh et al. [55] ©Copy right 2006, Radiation Effects and Defects in Solids.

of PI exhibited blister formation and nano-particles of different dimensions. They confirmed that these changes are caused mainly due to variations in local erosion rates. On exposure to oxygen plasma, various gaseous moieties like hydrogen, nitrogen, oxygen, and fluorine could be evolved from the surface region of polymer. The blisters on the surface of PI were reported to be ~1.4 μm and nano-particles formed in FEP and PI were, respectively, of dimensions ~50 to ~80 nm. By and large, their results showed that on exposure to plasma degradation PI was greater than in FEP under identical irradiation conditions. The obtained results were found to be consistent with estimated values of erosion yield for FEP and the PI.

5.6 XPS investigations

Samples of PI and FEP were exposed to plasma ions for about 1 h, followed by characterization using the XPS technique. The data were deconvoluted using origin with Gaussian fittings and FWHM ~1.6 eV as trial value for deconvolution. Binding energies were calibrated with respect to C 1s peak @ 285 eV (binding energy) in order to compensate for surface charging effects. **Tables 3** and **4** summarize the results. Peaks for both, PI and FEP, were assigned as per literature reports [37, 38, 71–77]. Peaks corresponding to C=C, C—C, and C—H, bonds in aromatic rings of

	Binding energy (eV) atomic oxygen		
Sample	Virgin	Irradiated	Assignment
PI	288.5	288.7	Shake up
	286.4	286.8	C—O—C
	285.5	285.7	Aromatic ring
	284.6	284.9	Aromatic structure
	283.8	284.2	Benzene ring
FEP	285	284.7	C—C or C—H
	–	286.3	C—O
	–	288.6	C=O
	290.8	290.6	—CF
	292.9	292.5	—CF$_2$
	295	294.5	—CF$_3$

Reproduced with permission Riyadh et al. [55] ©Copy right 2006, Radiation Effects and Defects in Solids.

Table 3.
C-1s binding energy values estimated from XPS peaks for virgin and plasma-treated PI and FEP.

Sample		Concentration %								
		C 1s	O1s	N1s	F1s	C:O	C:N	O:N	F:C	O:F
PI	Virgin atomic oxygen irradiated	68.23	29.3	2.6	—	2.6:1	26.2:1	11.3:1	—	—
		80.9	17.5	1.6	—	4.6:1	50.6:1	10.9:1	—	—
FEP	Virgin atomic oxygen irradiated	32.6	2.1	—	65.3	10:06	—	—	2.1	0:0.03:1
		37.5	19.2	—	43.3				1:15:1	0.44:1

Reproduced with permission Riyadh et al. [55] ©Copy right 2006, Radiation Effects and Defects in Solids.

Table 4.
Concentration of elementals, their ratios for virgin and plasma-treated PI and FEP.

PI appeared at 283.8, 284.6, and 285.5 eV for virgin PI and at 284.2, 284.9, and 285.7 eV for plasma-treated PI, respectively. Peak position shift was indicative of modification in imide structure that revealed the opening of the backbone benzene ring. For aromatic carbon, peaks emerged at 286.4 eV (virgin PI) and at 286.8 eV (plasma-treated) were assigned for its bonding with oxygen. Peak intensity of 286.8 eV was reduced by ∼50% compared to peak @ 286.4 eV. For virgin PI, peak at 288.5 and for plasma-treated PI peak @288.7 eV were assigned to a carbonyl group. They are also associated with esoteric and acidic functionality.

For virgin FEP, peaks at 285, 290.8, 292.9, and 295 eV were, respectively, assigned to C—H or C—C, —CF, —CF$_2$, and —CF$_3$ groups. Due to the large relative abundance of —CF$_2$; the presence of —CF and —CF$_3$was indistinguishable. For plasma-treated FEP, peaks @ 284.7, 286.3, 288.6, 290.6, 292.5, and 294.5 eV were, respectively, assigned to C—H or C—C, C—O, C=O, —CF, —CF$_2$, and —CF$_3$. For FEP, the intensity of —CF$_2$ peak was found to be reduced by ∼60% compared to virgin counterpart. The formation of free radical species with oxygen; led to the emergence of prominent oxidized carbon signals. After plasma treatment, positions of carbon in PI were shifted to higher energy and in FEP to the lower side. This can be explained on the basis of bond weakening in polymers. **Table 3** shows for FEP two additional oxygen-related moieties were formed due to oxygen replacement by fluorine. By and large, oxygen is comparatively less electronegative than fluorine. Therefore, in FEP binding energies were shifted in the lower regime for C-1s, whereas, in PI, the binding energy shift at the higher side was due to weak electronegativity of hydrogen as compared to oxygen. The imide structure was found to be modified by plasma by breaking aromatic rings [77] and inducing polarization near carbon sites. These processes showed an increase in the binding energy of C-1s in the XPS peaks of the PI.

The ratios of C:O, C:N, and O:N for PI and ratios of F:C, C:O, and O:F for FEP are provided in **Table 4**. The values were obtained after normalizing the intensity of each XPS peak with respect to the corresponding photoelectric cross-section. They were indicative of an increase in the concentration of carbon, whereas, reduction in O and N for plasma-treated samples. For hydrogen, XPS is relatively insensitive, and atomic oxygen and hydrogen are reactive to form water vapors out diffusing from a carbonaceous layer of polymer.

In PI, an increase in the ratio of C:N from 26.2:1 up to 50.6:1 for post-treated samples was attributed to atomic oxygen interactions with C—N bonds [77] and the formation of new C—O—N species. However, it was speculated that during plasma exposure, C—O—N bonding would be broken leading to new species like NOx on PI surface that might have been pumped out of the reactor. Desorption of water vapors and NO$_x$ species leads to a reduction in N and O with enhancement in C by plasma exposure. For FEP surfaces plasma treatment showed a significant increase in O, with a marginal increase in C including an appreciable reduction in F. This resulted in an increase in ratios of O:C and O:F on surfaces of plasma-treated FEP.

5.7 Vibration spectroscopy analysis

Figure 15 showed that, peaks in FTIR spectra recorded for virgin and plasma-treated PI (a) C—O stretching @ 1050–1250 cm^{-1}, (b) C—C backbone aromatic ring @ 1504 cm^{-1}, (c) N—H bend @ 1602 cm^{-1}, (d) carbonyl stretch @ 1724 cm^{-1}, and (e) O—H and N—H stretch @ 3080–2872 cm^{-1} [77]. They were similar to each other, except for small changes in intensities. Spectrum provided bulk information that was not so prominent in the surface region. Overall reduction in peak intensities was attributed to changes in the surface region mainly due to the erosion factor. **Figure 15** showed that peaks appeared at (a) 680–780 cm^{-1} were due to the presence of weaker band bending for C—F group, (b) @ 980 cm^{-1} due to —C—C—

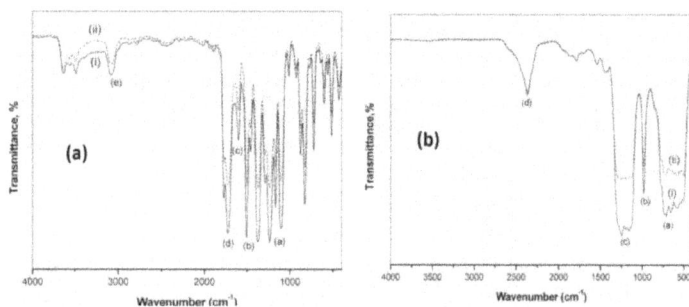

Figure 15.
FTIR recorded for (a) (i) virgin and (ii) plasma-treated PI, (b) (i) virgin and (ii) plasma-treated FEP (all @ 2 × 10^{17} ions cm^{-2}). Reproduced with permission Riyadh et al. [55] ©Copy right 2006, Radiation Effects and Defects in Solids.

stretch in side-chain C—CF$_3$ group, (c) @ 1100–1400 cm^{-1} for C—F coupled to C—C stretch, and (d) @2370 cm^{-1} for —CF$_2$ stretch [78–80].

For plasma-treated FEP, peak intensities were lowered with slight modifications in shape when compared with virgin FEP. Their results of FTIR supported the results of weight loss, SEM, and XPS for PI and FEP. The decrease in peak intensities revealed that the corresponding concentration of moieties had lowered due to atomic erosion from the polymer surface. From FTIR spectra of virgin and plasma-treated PI showed in **Figure 15**, it is clearly observed that the intensities of the absorption peaks at 1050–1250, 1504, 1602, 1724, and 3080–2872 cm^{-1} corresponding to C—O, C—C, N—H, C=O, and O—H, respectively, have decreased after exposure to atomic oxygen ions. Similarly, from the FTIR spectra of FEP shown in **Figure 15**, it is observed that the intensities of the C—F peaks at 680–780, 980, and 1100–1400 cm^{-1} have decreased after exposure to atomic oxygen ions. It is clear from these FTIR spectra that the bonds are more or less equally affected in FEP and PI after exposure to the atomic oxygen ions. However, the absorption peaks corresponding to O—H and N—H bonds showed a significant loss in the peak intensity. Their results, therefore, indicated that weight loss in PI was higher than in FEP. XPS analysis also revealed that C—O peak intensities in PI and intensities corresponding to fluorocarbon bonds in FEP have decreased, significantly, after exposure to plasma. Overall FTIR revealed that surface and sub-surface regions of PI and FEP were modified due to atomic erosion. Notably, the surface of FEP was reported to be more resistant to plasma attack. Riyadh et al. speculated that diffusion of fluorine into surface layers of PI may provide protection against atomic oxygen radiation degradation and have promised in space applications.

6. Synergetic effects: electrons + atomic oxygen

In this work, carried out by Majeed et al., samples of dimensions 15 mm × 15 mm × 50 μm were cut out from a PI sheet. The irradiation with electron beam was carried out by keeping samples inside a thin-walled polyethylene bottle mounted on the Faraday cup cage. From Race-track-microtron of Pune University, India, a pulsed electron beam of energy 1-MeV energy was obtained with pulse width ∼ 1.6 μs and pulse repetition rate 50 pps. Beam was scattered elastically by a thin tungsten foil of thickness ∼40 μm in order to obtain a uniform intensity on samples. The intensity of beam uniformity over a circular area of ∼40 mm was measured to be ∼5%. A current integrator, mounted in the system was used to measure electron flux incident on Faraday cup. Four different electron fluences 0.5, 1.0, 1.5 and 2.0 × 10^{15} e/cm^2 were chosen with constant flux. Thus, by varying irradiation time, electron-fluence was changed followed by exposure to plasma flux at different fluences of atomic oxygen ions. The experimentation was carried out in a separate setup. In this, reference samples were PI taken to be samples without electron irradiation.

For atomic oxygen plasma treatment, experiments were carried out in the in-house-developed, microwave-based Electron Cyclotron Resonance (ECR) plasma system as described in Section 5.1, **Figure 12**. The specific parameters such as energy and ion flux, in ECR plasma at different axial positions of the chamber. Chamber pressure variations were measured using a three-grid retarding field analyzer (RFA) as reported in our earlier publication [55]. For plasma treatment, samples were irradiated at an operating pressure of 8×10^{-3} mbar, at maximum energy position (10–14 eV) of oxygen ion with maximum flux ($\sim 5 \times 10^{13}$ ions cm^{-2} s^{-1}) with mean energy ~ 12 eV. Plasma was also characterized by a quadrupole mass spectrometer. They noted almost one order magnitude of variation at the higher side in the intensity of atomic oxygen ions than that of molecular oxygen over the entire energy range. Thus, they reported an abundance of molecular oxygen-less compared to atomic oxygen. Total fluence was varied by time variations in sample exposure, covering a range 5–20 $\times 10^{16}$ ions cm^{-2}.

Gravimetric measurements of virgin and radiation treated (electron + plasma) were performed using a microbalance with an accuracy of ± 1 µg.

Surface-wetting characteristics with de-ionized water were estimated by measuring contact angle "θ" in which measurements were performed with NRL-C. A. goniometer (RameHart. Inc. USA) with an optical protractor. Wettability measurements were immediately performed after removing samples from the plasma chamber. The wettability technique used was the sessile drop technique with 2 µl droplet volume. Three measurements were taken and values reported by Majeed et al. were the mean value of three measurements recorded for each sample. Measurement accuracy was ± 1 degree. The work of adhesion (WA) related to surface wettability was calculated which is [81] is given by: $W_A = \gamma_{LV} (1 + \cos\theta)$, where, LV is interfacial surface tension @ liquid/vapor interface, and θ is contact angle (LV = 72.8 mJ/m^2 for water).

Figure 16 shows variations in weight loss in % with plasma exposure time. Net weight loss in virgin PI was reported to be increased with plasma exposure time.

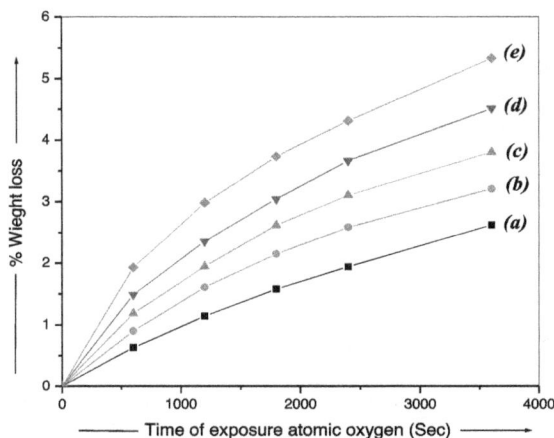

Figure 16.
Variations in % weight loss with plasma exposure for equivalent flux $\sim 5 \times 10^{13}$ ions cm^{-2} s^{-1}, for PI irradiated with 1 MeV electrons @ fluences: (a) 0 to (e) 20 $\times 10^{14}$ e/cm^2. Reproduced with permission Riyadh et al. [55] ©Copy right 2006, Radiation Effects and Defects in Solids.

Notably, for 1 MeV electron-irradiated PI rate of weight loss increased with increasing fluence of electron irradiation.

6.1 Investigations on atomic oxygen erosion

Erosion (volume per atom) yield quantification was carried out using the expression: $E_y = \Delta M / (\rho.A.\Phi.t)$, where, ΔM is mass variation, ρ, is density (1.42 gm cm^{-3}), A is sample area, ϕ is oxygen ion flux, and t is exposure time. The average erosion yield, Ey, was determined to be $\sim 7.05 \times 10^{-22}$ cm^3 atom^{-1} for virgin PI and $\sim 14.26 \times 10^{-22}$ cm^3 atom^{-1} for electron irradiated PI. **Table 5** displays values of Ey obtained for electron irradiated PI @ different fluence.

Data revealed that, average erosion yield after exposing PI to electrons has increased by a factor of 2 as compared to virgin PI which was indicative of PI upon electron irradiation becomes more susceptible to plasma erosion.

In order to reveal the nature of radiation-induced surface modifications; hydrophilic characteristics of polymer surface were investigated by measuring wettability. Wettability depends upon surface roughness and surface chemical composition. Change in surface properties was, in turn, reflected in modification in measured contact angle. **Figure 17** shows variations in contact angle plasma exposure time, for PI irradiated with 1 MeV electrons @different fluences. They noted a reduction in

Fluence of 1 MeV electron ($\times 10^{14}$ e/cm^2)	Maximum erosion yield ($\times 10^{-22}$ cm^3 atom^{-1}) (accuracy 1%)
0 (virgin)	7.28
5	8.92
10	10.56
15	12.53
20	14.81

Table 5.
Maximum erosion yield (E_y) for PI irradiated @ different electron fluences with subsequent exposure to oxygen plasma flux@ $\sim 5 \times 10^{13}$ ions cm^{-2} s^{-1}. Reproduced with permission Riyadh et al. [55] ©Copy right 2006, Radiation Effects and Defects in Solids.

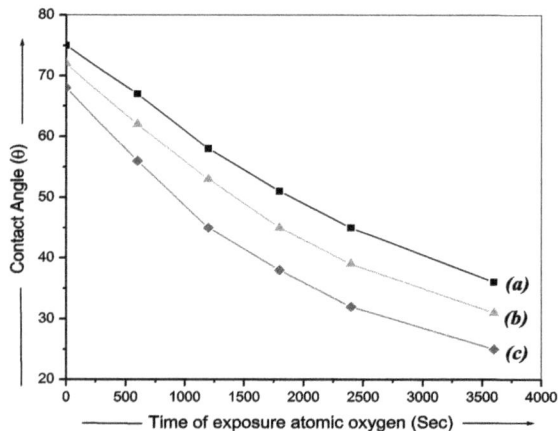

Figure 17.
Contact angle variations with plasma exposure (@ $\sim 5 \times 10^{13}$ ions cm^{-2} s^{-1} flux) for PI irradiated with 1 MeV electrons with fluences: (a) 0 (virgin), (b) 10 $\times 10^{14}$ e/cm^2, and (c) 20 $\times 10^{14}$ e/cm^2. Reproduced with permission Riyadh et al. [55] ©Copy right 2006, Radiation Effects and Defects in Solids.

| Time of exposure (s) | Virgin PI samples | | 1 MeV electrons irradiated samples at fluence | | | |
| | | | $(10 \times 10^{14}$ e/cm$^2)$ | | $(20 \times 10^{14}$ e/cm$^2)$ | |
	Contact angle "θ" (°)	Work of adhesion (W$_A$) mJ/m^2	Contact angle "θ" (°)	Work of adhesion (W$_A$) mJ/m^2	Contact angle "θ" (°)	Work of adhesion (W$_A$) mJ/m^2
0	75	91.6 ± 1.3	72	95.3 ± 1.2	68	100.1 ± 1.2
600	67	101.2 ± 1.2	62	107 ± 1.1	56	113.5 ± 1.1
1200	58	111.4 ± 1.1	53	116.6 ± 1.0	45	124.3 ± 0.9
1800	51	118.6 ± 1.0	45	124.3 ± 0.9	38	130.2 ± 0.8
2400	45	124.3 ± 0.9	39	129.4 ± 0.8	32	134.5 ± 0.7
3600	36	131.7 ± 0.7	31	135.2 ± 0.7	25	138.8 ± 0.5

Reproduced with permission Riyadh et al. [55] ©Copy right 2006, Radiation Effects and Defects in Solids.

Table 6.
Data for θ and W$_A$ for virgin and electron irradiated PI before and after plasma exposure for different periods of time.

water contact angle, θ, to a larger extent as compared to virgin PI. Corresponding W$_A$ was calculated and displayed in **Table 6**.

The obtained data was indicative of θ change with W$_A$ for electron irradiated PI which was observed to be much higher than that for virgin reference. Significant reduction in θ with improved wettability was mainly attributed to electron irradiation-induced surface modifications.

Morphological analysis was carried out by investigating changes in surface roughness. In **Figure 18(a)** and **(b)**, SEM micrographs, comparatively show surface-morphologies of PI exposed to plasma; with and without electron

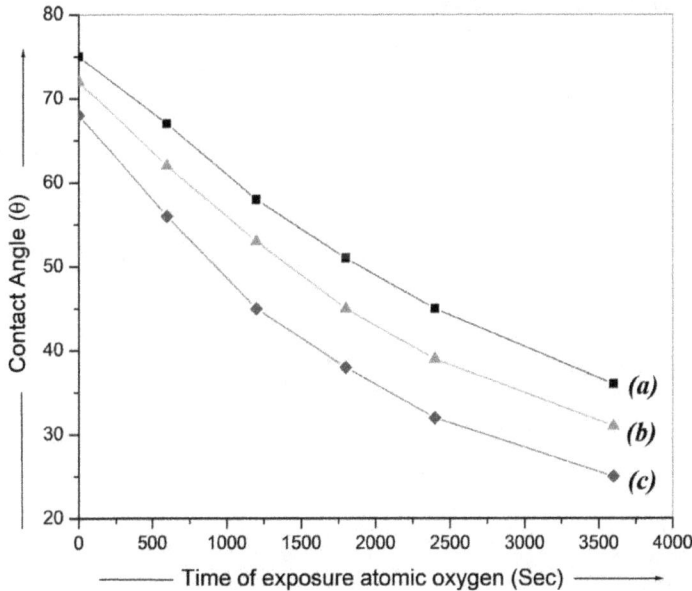

Figure 18.
SEM micrographs: (a-i) virgin, (a-ii) plasma treated (@ 2 × 10^{17} ions cm^{-2}), (b-i) electron irradiated, (b-ii) combined (electron + plasma treated) @ 2 × 10^{17} ions cm^{-2}. Electron fluence @ 20 × 10^{14} e/cm^2 for both samples in (b). Reproduced with permission [52] Copy right 2006, Radiation Effects and Defects in Solids.

pre-irradiation. Although surface roughness was noted to be enhanced marginally when pre-irradiated alone (b-i) or oxygen alone (a-ii). However, for electron irradiated surfaces (@ 20×10^{14} e/cm^2) that were subjected to plasma (@ 2×10^{17} ions cm^{-2}) a drastic change in surface roughness was observed, as seen in (b-ii). The surface texture of electron-irradiated PI that was subsequently exposed to plasma was observed to consist of globules, cavities, and blisters of various sizes. This may be attributed to synergetic electron-irradiation and plasma-induced surface erosion in PI. Majeed et al. reported that defects introduced in the surface region by electron irradiation indent roughness which became more susceptible to oxidative erosion by plasma treatment. The chain scission process seems to be responsible for causing surface degassing from the samples. Due to the coagulation of emerging molecular species, blisters were formed. Nonuniform degassing may lead to significant surface roughness. They reported secondary effects too such as chain scission might have re-crosslinked leading to a subsequent change in surface free volume, which resulted in the imperfect surface after radiation [52] treatment.

As discussed earlier, polymer on encountering energetic radiations breaks in chain and link as well simultaneously. However, chain scission or cross-linking depends upon the type of polymer, radiation, and dose rate. As reported in earlier literature [52] it appears that MeV electrons can effectively produce cross-linking in PI. Moreover, the dose rate of electrons is considerably high. In such cases radiation density induced zones nucleates in reactive free radicals that expand and get overlapped resulting in cross-linking of free radicals. Moreover, the dominance of scission or cross-linking also depends upon the extent of radiation-induced oxidation.

The combined effect of radiation-induced changes was further analyzed by Majeed et al., using vibrational i.e. FTIR spectroscopy. **Figure 19** shows recorded FTIR spectra for and electron-irradiated PI before and after exposure plasma exposure. Four spectra are displayed in the figure. Basic spectrum corresponds to virgin i.e. profile (i), (ii) plasma-treated, (iii) electron irradiated, (iv) combined electrons + oxygen plasma. Intensities in (i), can be compared with those in other spectra. Prominent absorption was observed corresponding to (a) C—O str. @ 1050–1250 cm^{-1} (b) CO—N—CO cyclic imide appeared @1376–1390 (=C—N str.), (c) C—C backbone aromatic ring @ 1504 cm^{-1}, (d) N—H bend. and str., respectively, @1602 cm^{-1} and 3486 cm^{-1}, (e) carbonyl str. @ 1724–1780 cm^{-1}, (f) C—H str. @ 3085 cm^{-1}, and (g) hydroxyl —O—H @ 3630 cm^{-1} [66, 82, 83]. By and large, Majeed et al., noted a marginal reduction in intensities of almost all emerged bands

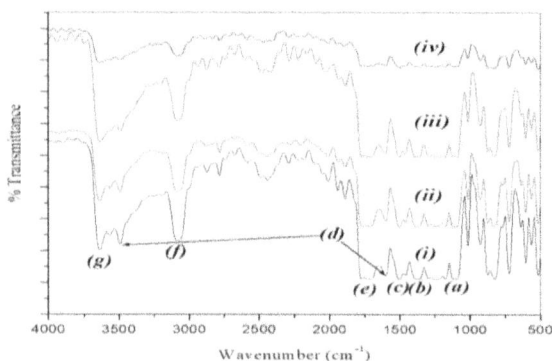

Figure 19.
Recorded FTIR spectra of PI for (i) virgin, (ii) plasma treated @ 2×10^{17} ions cm^{-2}, (iii) pre-electron-irradiated @ 20×10^{14} e/cm^2, and (iv) electrons (@ 20×10^{14} e/cm^2 + plasma (@ 2×10^{17} ions cm^{-2}) treated. (a) to (g) indicate assigned functional groups described in the text. Reproduced with permission Riyadh et al. [55] ©Copy right 2006, Radiation Effects and Defects in Solids.

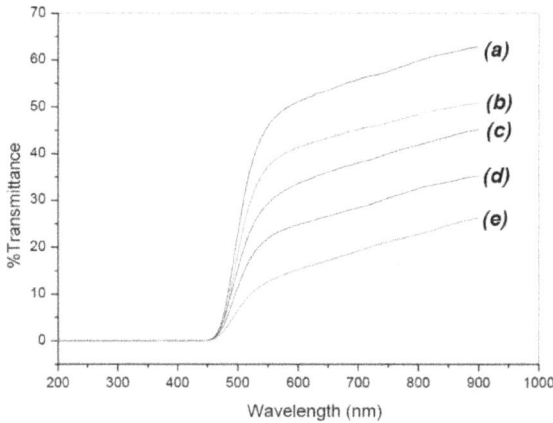

Figure 20.
UV-visible transmission spectra of (a) virgin, (b) plasma treated, (c) electron irradiated + plasma treated (plasma @ 1 × 10^{17} ions cm^{-2}) (d) plasma treated, (e) electrons + plasma treated (plasma @ 2 × 10^{17} ions cm^{-2}). Reproduced with permission Riyadh et al. [55] ©Copy right 2006, Radiation Effects and Defects in Solids.

on exposure of the thin film to plasma, whereas, electron alone affected to a lesser extent showing shallow changes in absorption peaks. In contrast, IR absorption bands were extensively reduced for combined electrons and plasma effects. This showed electron irradiation alone does not extensively contribute to chain scission or cross-linking, whereas, plasma irradiation is capable of producing chain scission as well as cross-linking by the reactive oxidation process. This process got further prominent if the ionizing electron irradiation has created defect centers. As reported earlier, atomic oxygen reacts with C—N species which is the weakest bond in PI, and establishes C—O—N bonding.

Figure 20 showed UV visible transmission spectra of PI @ different radiation conditions. A combination of two doses (electron + plasma) was presented. The plots showed that the value of % transmission was observed to be decreased when PI got treated with electrons + plasma combinedly as compared to plasma alone. Thus, synergetic effects were operational by electrons irradiation causing surface erosion, and apparently plasma treatment effectively enhanced the roughness of polymer films. Such a rough surface was responsible to enhance light scattering and thus reducing the overall transmission property of the sample.

7. Ion track technology: production parameters

7.1 Techniques of track etching

In a study carried out by Alegaonkar et al., common polymer foils were used such as polyethylene terephthalate (PET), PI (PI), polycarbonate (PC), and silicone rubber, however, due to space crunch, herein, we have presented only the results of 10–20 μm thick PET and PI films (Goodfellows Ltd., Russian). These films were irradiated with 300–500 MeV Ar, Kr, Xe ions over the fluence 1–500 × 10^6 cm^{-2}@ heavy ion accelerator @JINR Dubna (Russia) or @ HMI Berlin (Germany).

By and large, the choice of the projectile and its energy provides magnitude and distribution of transferred energy density along the ion track trail, which consequently determines the track-to-bulk etching rate. It also influences the shape of emerging pores called etched ion tracks (post-etching). For PET irradiated under certain conditions and etched in a typical condition yields rather cylindrical tracks, whereas, for PI conical (or hyperbolical if etched from both sides) had been reported. The angle of cone, χ, was observed to be increased with reducing an atomic number of projectile ion, $\chi \sim 20°$ for PI. Post latent track etching of PET films was carried out using 3 mole/l NaOH at 45°C. Systematic observations on the etching track shapes were made in a regular time interval using ion transmission spectrometry (ITS) [84]. ITS technique records the energy spectrum of nearly monoenergetic (5.49 MeV) α particles from a 40 kBq ^{241}Am α source, transmitted through the etched films. With sequential etching, when for the first time α particles get transmitted through the etching film without any energy loss it is indicative of etching breakthrough can be achieved. One can calculate track radius, provided that one knows the areal track density [85]. The tracks etching has been done up to diameters in a range of $\sim$100 nm and a few μm.

7.2 Production of metallic tracks: chemical deposition

By deposition of Ag, Cu, and Ni, we have carried out the production of conducting etched ion tracks. We adopted the chemical or electrodeless deposition (ELD) technique in order to deposit metal within the etched tracks [86]. Particularly for Cu and Ni deposition, we used commercially available solutions from M/s Doduco Ltd. Whereas, for deposition of Ag, we followed the reported procedure of St. Gobain (Brockhaus, 1895; Th Pone, 1955), due to the green approach. Deposition of Cu and Ag was carried out at 24°C, and Ni @ $\sim$88°C. Martin et al. [86] reported a strong surface-near reduction of the inner track diameter including the closure of track, however, we have not observed any such thing.

7.3 Nucleation centers

7.3.1 Chemical activation

In order to achieve durable metal layers on polymeric substrates, it requires the existence of nucleation centers on polymer surfaces where the metal atoms could be deposited. For polymers used herein, for metal precipitation and nucleation the areal density intrinsic surface defects were too low for the formation of a continuous metallic layer. This led to the emergence of discontinuous metal tracks as seen

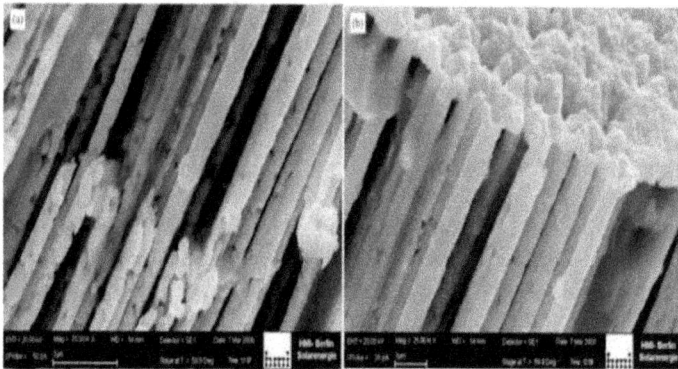

Figure 21.
Comparison of SEM micrographs for (a) intermittent and (b) continuous ELD-deposited tracks of Ag in PET. In this, image (a) shows preparation of deposition without activation, and image (b) displays the addition of chemical activation centers. Reproduced with permission Fink et al. [6] ©Copy right 2004, Radiation Effects and Defects in Solids.

in **Figure 21(a)**. The creation of additional nucleation centers was observed to be generating continuous track tubules as seen in **Figure 21(b)**. One can achieve this by surface activation process either chemically by bonding suitable metal atoms (e.g. Sn, Pd) or physically by creating dangling surface states by laser or ion irradiation.

We have carried out resistivity measurements along the track axis to assess the quality of emerging track deposition. This was carried out by contacting the front and rear sides of the etched track polymer films using Au electrodes under constant gentle pressure. Initially, high-track resistivities were registered. In order to determine very low track resistivities, some polymer films were irradiated through mask pinholes to reduce the areal density of tracks (which were counted).

Figure 22 shows the dependence of Cu-filled track quality with activation time that revealed a necessity to introduce a sufficient number of nucleation centers. For non-activated samples, the resistivity was (not shown here) closely resembled samples of 1s activation time (**Figure 22**). Analysis exclusively revealed that the metal was deposited onto intrinsic defects. From **Figure 22**, it can be seen that after the exposure for a few seconds the process of activation improves the track quality by orders of magnitude. On saturation of density of nucleation centers, exposure of polymer film to the activation solution is ineffective. Thus, the quality of metallic track-tubules is correlated to the roughness of their inner surfaces. During the initial track, tubule growth surface area is larger and nucleation centers were less available. For smooth inner tubule walls with high conductivity, a high nucleation center density is a basic requirement. Moreover, the roughness of inner walls of the etched track, depending on the etching conditions such as etchant concentration; the smoothness of the outer tubule surfaces can be determined) and is of utmost importance. This is demonstrated by comparing the quality of metal-filled tubules in PET and PI prepared under identical conditions. We noted the filled track tubules in PI were better due to the greater smoothness of both inner and outer tubule walls.

7.3.2 Formation of nucleation centers by ion irradiation

In addition to chemical activation, another channel is ion irradiation. The underlying mechanism is to activate ring structures in polymer by electronic energy transfer so that oxygen can be attached. This will facilitate metal addition. The nucleation center formation experimentation for 30–300 keV He+ to Xe+ ions was performed that resulted in polymer surface activation @fluences of 10^{13} cm^{-2} and

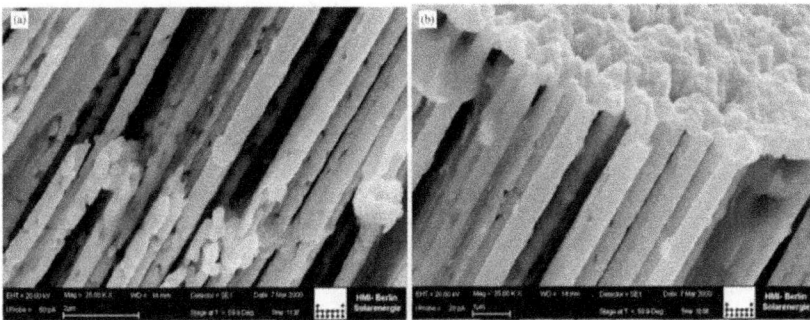

Figure 22.
Variation in resistivity of tracks in PE and PI films with embedded Cu material along the surface normal to the film (i.e. along the axis of ion track). Conditions for chemical activation and ELD deposition: 5 min ELD solution exposure. Reproduced with permission [6] Copyright 2006, Radiation Effects and Defects in Solids.

was noted to be independent of projectile ion. No saturation in surface activation was observed due to irradiation.

Morphological investigation revealed that metal layers deposited onto ion-activated polymers were appeared to be smoother compared to the chemically activated nucleation process. They concluded ion-induced activation took place locally which enabled to tailoring of the metal deposition along the axis of the track. It means for a cylindrical metal contact embedded in tracks can be produced on both sides of a track for sensor application.

7.3.3 Characterizations of tubule

For metals, ELD deposition took place in track tubules after a system-dependent specific incubation time (Ni ~ 5 s, Ag ~ 1 min, Cu ~ 11.2 min). However, at higher temperatures, the track deposition and tubule formation speed were observed to be fast [8]. **Figure 23** via-a-vis compares the rate of deposition of ELD at room temperature within the etched ion tracks with unirradiated polymer surfaces. However, within the tubule, metal layers started growing ~2 times earlier than that on pristine surfaces which were attributed to the higher activation density of etched tracks. The irradiation-induced damage at the inner walls of the track were bonded the sensitizers better than inert pristine surfaces. For both tubules and pristine surfaces deposition showed a steadily decreasing deposition trend with an increase in time. This was attributed to exhaust of available solutions and the complete filling up of narrow tracks. No significant difference was observed for pristine surfaces and tubule walls of any radius, which that demonstrated, the diffusion speed of ELD solution within very narrow tracks has not a decisive parameter. A dramatic change in their electric conductivity was noted for increased metallization of filled tracked films. During the initial stage of deposition, the fall in resistivity was seen by orders

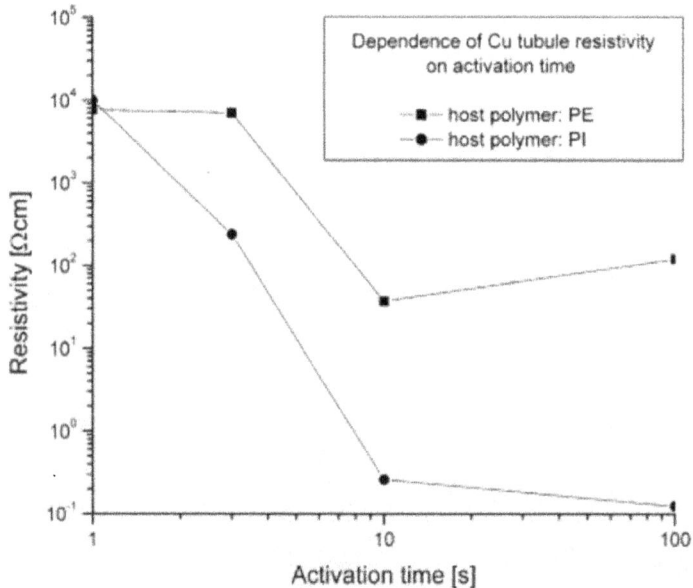

Figure 23.
Metallic layers deposition and growth in PI track by ELD. Comparison for deposition in pristine polymer films and growth at the interior of etched tracks, for Cu and Ni deposition. Reproduced with permission [6] Copyright 2006, Radiation Effects and Defects in Solids.

of magnitude from dielectric to semiconducting to conducting regime with the gradual development of continuous metal layer. Thus, further reduction in resistivity was determined by the increase in layer thickness. Inside the tracks, the resistivity was observed to be improved faster than on the surface of the thin film. This observation was consistent with the rise in the growth of metal layers within the tracks.

By and large, we were able to prove by SEM, conductivity studies, and ITS measurements that metallic tubules tracks were continuous and hollow throughout the track height of the polymer film. However, measurements of Rutherford backscattering revealed marginal variation in thickness of tubule wall along the axis of track.

7.4 Thermal stability of etched tracks and metallic nanotubules

There is number of proposals for ion track applications that required high-temperature stability either during device preparation or actual implementation, for such purpose behavior of the tracks and their embedded structures must be known. As a first examination of this kind is already reported in reference [87]. According to reported findings, etched tracks of PE initially gots welled a bit upon heating due to release of moisture (received from the etchant) and subsequently became thin due to onset flow of glassy polymer, and towards the end got widen again swiftly due to carbonization. As a result, silver-filled track tubules appeared to be lost in their parallel alignment above the glass transition temperature (T_g) so that their transmission rapidly decreased upon annealing. However, for PI, developed ion tracks were observed to be maintained their shapes up to ~450°C, and thereafter, transformed with the onset of carbonization. They were observed to be increased in their width. Silver-filled PI tracks were observed to be led to a stable structure >600°C.

8. Applications of ion tracks

Above discussion clarifies the role and application of latent (as-implanted ion tracks) and etched tracks. Broadly speaking, latent tracks are emerged fundamentally due to the deposition of high energy (MeV to GeV) inside a tiny volume element called as the core of the ion track which has volume 10^{-15}–10^{-14} cm^3. Moreover, within an impulsive time, about 10^{-17}–10^{-15} s interaction by swift heavy ions, these exceptional transient conditions, led to a dramatic variation within materials. These changes are chemical and structural in origin that accompanies heat and pressure impulse. Primary damage such as transient breaking of all bonds is recovered during generated annealing phase followed immediately after the ion impact, named, thermal spike in about 10^{-12}–10^{-11} s. However, a number of irreversible changes remained within the damaged cylindrical zone. Thus, such zone is altered with characteristic changes like accumulation of high density of radicals, excess carbonaceous clusters, new phases (SiC in polysilane), amorphization, and modification in inherent free volume. As a result, several macroscopic properties of the irradiated polymer could get modified like permeability, refraction index, electric conductivity, thermal properties, etc.

For such damaged zone i.e. latent tracks, there are four major strategies showed up: (1) exploiting modified transport properties along the tracks, (2) precipitation of metallic atoms or clusters along the damaged zone, (3) exploiting chemical changes in polymer, and (4) using phase transition characteristics induced by ion irradiation [88]. Hitherto insignificant work has been carried out in these fields. In general, ion-irradiated polymer films carrying latent tracks can be used as a seal in order to protect sensitive zones against penetration of ambient dust and moisture; maintaining equilibrium pressure and exchange of gases with ambiance. There are very few reports presenting proof of concept like doping of latent tracks for electronic application [85], SiC needles [88, 89] for AFM cantilevers, ion-induced conducting nanowires in diamond [90], or fullerite [88] for field emission displays are demonstrated.

8.1 Manipulations with etched tracks: flat flexible devices

Dissolving latent track material by suitable agents, termed as an etching process, led to the formation of the etched track. For this, there is a combo pack of projectile selection, targeted polymer, etchant, and etching conditions. Further, etched tracks can be manipulated in the desired shape like cylindrical, conical, hyperbolic, transmittent, non-transmittent. In principle, etched ion track can be filled by a number of material, and embedded assembly of material can be assembled as bulky wires, termed as, fibers or fibrils or tubules. In this, metal could be dispersed discontinuously in the form of nanoparticles along with the track geometry. The initial step could be realized by embedding matter within etched tracks to transport material clusters towards the required position which could be accomplished by dissolving material of interest in a suitable aqueous or non-aqueous solution followed by penetration of the solution into the etched tracks by capillarity action. By allowing liquid to evaporate within the tracks, followed by dissolution of matter could result in a supersaturated state of metal particles which led to precipitate onto form tubule. There was number of reported tubules like KCl, C_{60}, polymethyl methacrylate (PMMA), dyes, etc. The thickness of the tubule depends, empirically, depends upon the amount of matter dissolved in a solvent and transported into the ion track.

The class of such deposition is called as chemical or electrodeless (ELD) deposition technique [91]. In this, high degree of supersaturation of certain metals like Cu, Ag, Au, Ni, etc., or metal chalcogenide like CdS, PbS, Cu_xS, ZnS, CdSe, ZnSe, CdTe, CuInSe2, etc., could be achieved by dissolving them in a suitable agent such as NH_3, ethylenediamine (EN), nitrilotriacetate (NTA), thiourea (TU), and others. They precipitate heterogeneously to generate tubules of these materials in etched ion tracks which possess nucleation centers activated chemically or physically or chemically as discussed earlier.

Further, colloidal possess specific optical and electronic properties like high charge storage capacity, fluorescence, polarization, etc., blending colloidal with etched track could lead to new interesting applications like colloidal of TiO_2, and $LiNbO_3$ (20 nm: ball milled) to form nanotubules in etched tracks for optical sensors. Moreover, solver colloidal of conducting silver paste was used to deposit etched tracks that yielded conducting micro/nanowires for etched ion-track with diameter five times larger than silver particles. In etched ion tracks, it is possible to deposit material by carrying out chemical reactions like the formation of photosensitive AgBr, by allowing chemical reaction between $AgNO_3$ and NaBr within etched ion tracks in which tracks were acted as microreactors. However, the hydrophilic nature of tracks puts requirements on penetrants to have wetting characteristics. For hydrophobic solvents such as liquid metals, one can use a pressure injection route to enter solvent forcedly into the nanopores. Eventually, this could be useful in modifying the wettability characteristics of etched track walls by suitable deposits. We observed that liquid Pb/Sn solder material could readily be penetrated into etched ion tracks, therefore, can be used for contacting ion-track-based devices for Cu-filled tracks.

Further, galvanic deposition is another deposition technique that was reportedly used to coat conducting matter within etched tracks. In general, etched microporous polymeric films could be connected to the cathode such that the deposited metal or conducting polymer grows throughout etched tracks generating non-porous rods. However, a major disadvantage in all track manipulation techniques using solvents was noted to be the formation of ionic radical e.g. Ag tubules in PI possess a weak halo of Na^+ (etchant) and Ag^+ (ELD) ions contributing weak ionic bulk conductivity. Etched ion tracks can also be deposited by gas-phase pyrolysis type reactions, permeating gas to transform solid residues such as carbon or silicon, via pyrolytic reactions. However, such reactions required high reaction temperatures, and etched tracks in PI, mica, and SiO_2 were found to be well-suited candidates. Evaporation also led to a limited deposition near the entrance of the ion track and having a demerit of limited depth that can be reached in narrow tracks by evaporation [92]. Semiconducting material filling in ultra-small dimensional tracks, exhibited diode characteristics with quantum effects such as resonant electron tunneling in which current/voltage characteristics exhibited characteristic local maxima and minima over the applied voltage range. Capacitive coupling of such semiconducting ion track wires led to the formation of field-effect transistor-like structures. Especially, surface-modified metal ion track nanotubules were reported to be used to control electrolytic permissivity [93]. Moreover, such nanotubules grafted with gels were performed like switches for penetrating liquids [94]. Au nanotubules were used as miniaturized electrolytic sensors for detecting oxygen concentration. They were also used for in vivo long-time drug storage [95]. TiO_2 nanotubules were found to be advantageous in Li batteries [96] and fullerene tubules were reported to be effective

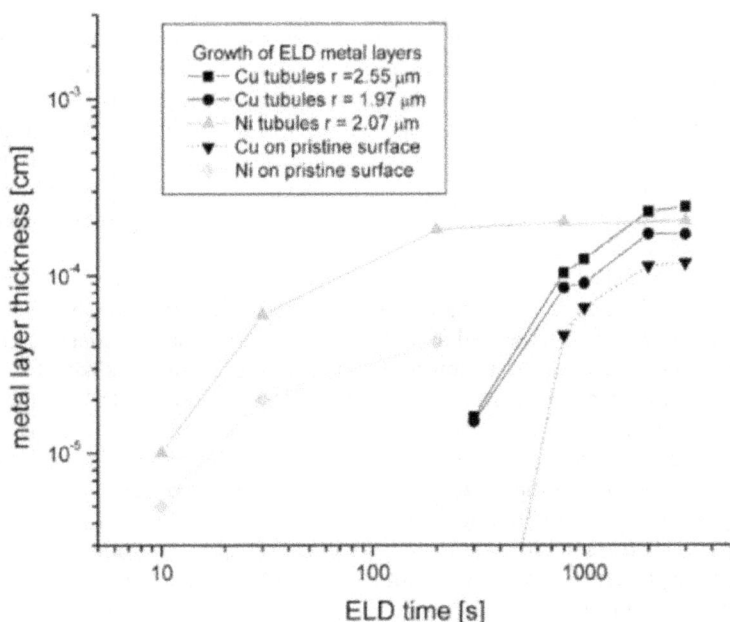

Figure 24.
Etched ion-track-based magnets and transformers, Left panel (a) prototype magnet with dimensions 6 × 7 × 0:02 mm³, deposition masks (above) and production steps (below); (b) comparison of input (bottom) and output (top) signals of the transformer prototype, for rectangular and sinusoidal input at different frequencies. The reasonable functional form is recorded in MHz frequency range (upper graph), but the signal transformation is possible up to the near-GHz range when gradually a phase lag emerges between input and output signals; (c) first miniaturized ion-track-based transformer of 1:2 × 0:25 × 0:01 mm³ size. Reproduced with permission Fink et al. [8] ©Copy right 2006, Radiation Effects and Defects in Solids.

temperature, pressure, and humidity sensors [97]. Layer-by-layer galvanic deposition of distinct materials led to axially structured elements such as Cu and Se [98, 99], or Ni and CdSe/CdTe [100] showed nano-diode characteristics, whereas, sequential galvanic deposition of different ferromagnetic materials led to giant magnetic resonance, GMR [101]. Layer-by-layer chemical deposition of semiconductor/metal/semiconductor led to the formation of cylindrical diodes, and metal/insulator/metal structures resulted in concentric condensers [102].

Architecting copper metal wires into etched ion tracks with sequential evaporation of metal contact stripes onto the surface of microporous polymer film led to the formation of micro-inductors (**Figure 24a**) and micro-transformers. Correspondingly, masks used for evaporation and intermediate production stages of a prototype magnet are displayed in **Figure 24a**. Corresponding test transformers show good operation up to ~0.5 GHz (**Figure 24b**), with inductances per winding of 10^{-4} H @ 10 Hz and 2.5×10^{-7} H @10^4 and 5×10^8 Hz with Q-factor approaching up to ~10. The miniaturized ion-track-based transformer is shown in **Figure 24c**. In a similar fashion combining lithography and etched tracks, miniaturized magnetic field sensors based on magneto-resistive ion tracks were produced by another research group [101]. First, ion-track-based sensors were developed especially for oxygen, as some biochemically important materials to be detected reacts with the enzyme glucose oxidase towards H_2O_2 which was transformed to water by anodic oxidation. Also, temperature, pressure, and humidity sensors were developed with C_{60} as a sensor material [97, 103].

9. Conclusions

In conclusion; the modification in various electrical and physicochemical properties of polyimide is shown to be induced by novel methods; without changing the major chemistry of the polymer. Most of the reported investigations are related to the irradiation effects of 50 μm thick, space quality film of polyimide by high energy electrons, gamma rays from Co-60, swift heavy ions, and low energy ions of atomic oxygen generated from the plasma source. Extracts of experimental details and the results have been presented in brief and are comprised in five separate sections (Sections 2–7) of the chapter.

The effects of electron irradiation on the dielectric properties were compared with those obtained from diffusing certain elements such as fluorine, boron, and phosphorous by radiation assisted diffusion process. The dielectric constant and the refractive index are shown to be lowered by irradiating the samples with 1 MeV electrons. It was seen to be further lowered when PI was doped with boron and fluorine even into a thin (3 μm) surface layer on both sides of the film. However, similar diffusion of phosphorous has been reported to increase the dielectric constant from 3.15 to 4.5. In these studies, the RBS technique was used for depth profiling wherein XPS for finding the chemical changes occurring in the surface layers. The variations in the dielectric constant and dielectric loss obtained with Co-60 gamma rays were similar to those obtained with 1 MeV electrons at a fluence of 10^{15} e/cm^2. These results are important in view of using 50 μm thick PI in space applications since the space vehicle is exposed to different kinds of radiations where lower values of dielectric constant are preferred. It is noteworthy that the diffusion of phosphorus or boron and fluorine was carried out at room temperature and the phenomenon of the tailoring of the dielectric characteristics of PI was found to be surface sensitive.

There are reports wherein the mechanical properties of PI were found sensitive to the radiations. Noticeable enhancement in the tensile strength was observed when PI was irradiated with 1MeV electrons and this effect is attributed to the radiation-induced cross-linking. In addition, the electron beam-induced surface roughening was found to enhance the hydrophilicity of PI surface. In conclusion, it is proved that high-energy electron irradiation can be used to tailor the bulk and surface properties of PI.

Being a space-compatible polymer, PI has a special role to play in supersonic aircraft and in space vehicles. However, there is always a concern about its possible degradation on exposure to the atomic oxygen species during its passage in LEO. The paper discusses the various effects of AO, generated in a home-built Electron Cyclotron Resonance plasma system, on the surface erosion properties. It is noticed that degree of the surface degradation in FEP is lower, by a factor of five than in PI. The synergetic effects of 1 MeV electron irradiation and exposure to AO irradiation of PI have been discussed in detail. Electron irradiation has been found to be more detrimental to the surface erosion behavior by AO. The surface defects arising from chain-scission and cross-linking in PI by electrons are set responsible for the increase in the erosion yield.

Our earlier studies related to the production and behavior of metal nanostructures by chemically filling the etched ion tracks (produced by swift high energy ions) in PI have been elaborated. The results have proved the strength of the method for future devices. These reports indicate that the Etched Ion Tracks, in combination with lithography, opens up the possibilities for creating novel micro-structures in PI which is inaccessible by other techniques.

Acknowledgements

The authors are thankful to the Director, ISRO-DRDO, SPPU Cell for providing an opportunity to carry out work. They are thankful to the Director, UGC-DAE-CSR for providing access to their facilities, as and when required. PSA and VNB are thankful to ISRO, German Academic Exchange Program (DAAD), and involved scientists for giving an opportunity to carry out work at Hahn-Meitner-Institute, Berlin, Germany. The authors are thankful to Dr. Ashwini P. Alegaonkar for compiling and typesetting the chapter.

Author details

Prashant S. Alegaonkar[1*], Vasant N. Bhoraskar[2] and Sudha V. Bhoraskar[2]

1 Department of Physics, School of Basics Sciences, Central University of Punjab, Bathinda, Punjab, India

2 Department of Physics, Savitribai Phule Pune University, Pune, Maharashtra, India

*Address all correspondence to: prashant.alegaonkar@gmail.com

IntechOpen

References

[1] Iredale R, Ward C, Hamerton I. Modern advances in bismaleimide resin technology: A 21st century perspective on the chemistry of addition polyimides. Progress in Polymer Science. 2017;**69**: 1-21. DOI: 10.1016/j.progpolymsci. 2016.12.002

[2] Murphy C. Polyimides: Synthesis, Applications and Research. New York: Nova Science Publishers; 2016

[3] Ghosh MK, Mittal KL. Polyimides: Fundamentals and Applications. New York: Marcel Dekker; 1996

[4] Tribble AC. The Space Environment: Implementation for Spacecraft Design. Princeton, NJ: Priceton University Press; 1995

[5] Grossman E, Gouzman I. Space environment effects on polymers in low earth orbit. Nuclear Instruments and Methods in Physics Research Section B. 2003;**208**:48. DOI: 10.1016/S0168-583X (03)00640-2

[6] Fink D, Petrov A, Rao V, Wilhelm M, Demyanov S, Szimkowiak P, et al. Production parameters for the formation of metallic nanotubules in etched tracks. Radiation Measurements. 2003;**36**(1-6):751-755. DOI: 10.1016/ S1350-4487(03)00209-9

[7] Martin C. Nanomaterials: A membrane-based synthetic approach. Science. 1994;**266**(5193):1961-1966. DOI: 10.1126/science.266.5193.1961

[8] Fink D, Alegaonkar PS, Petrov AV, Berdinsky AS, Rao V, Müller M, et al. The emergence of new ion tract applications. Radiation Measurements. 2003;**36**(1-6):605-609

[9] Xu G, Gryte C, Nowick A, Li S, Pak Y, Greenbaum SG. Dielectric relaxation and deuteron NMR of water in polyimide films. Journal of Applied Physics. 1989;**66**(11):5290-5296. DOI: 10.1063/1.343719

[10] McCrum NG, Read BE, Williams G. Anelastic and Dielectric Effects in Polymeric Solids. Vol. 126. New York: Wiley; 1967

[11] Cha H, Hedrick J, DiPietro R, Blume T, Beyers R, Yoon DY. Structures and dielectric properties of thin polyimide films with nano-foam morphology. Applied Physics Letters. 1996;**68**(14): 1930-1932. DOI: 10.1063/1.115629

[12] Maruo Y, Sasaki S, Tamamura T. Change in refractive index and in chemical state of electron beam irradiated fluorinated polyimide films. Journal of Vacuum Science & Technology A: Vacuum, Surfaces, and Films. 1995;**13**(6): 2758-2763. DOI: 10.1116/1.579701

[13] Todd M, Shi F. Characterizing the interphase dielectric constant of polymer composite materials: Effect of chemical coupling agents. Journal of Applied Physics. 2003;**94**(7):4551-4557. DOI: 10.1063/1.1604961

[14] Hougham G, Tesoro G, Shaw J. Synthesis and properties of highly fluorinated polyimides. Macromolecules. 1994;**27**(13): 3642-3649. DOI: 10.1021/ma00091a028

[15] Hougham G, Tesoro G, Viehbeck A. Influence of free volume change on the relative permittivity and refractive index in fluoropolyimides. Macromolecules. 1996;**29**(10):3453-3456

[16] Hougham G, Tesoro G, Viehbeck A, Chapple-Sokol J. Polarization effects of fluorine on the relative permittivity in polyimides. Macromolecules. 1994; **27**(21):5964-5971

[17] Shih D, Yeh H, Paraszczak J, Lewis J, Graham W, Nunes S, et al. Factors affecting the interconnection resistance

and yield in multilayer polyimide/ copper structures. IEEE Transactions on Components, Hybrids, and Manufacturing Technology. 1993;**16**(1): 74-88. DOI: 10.1109/33.214864

[18] Laghari J, Hammoud A. A brief survey of radiation effects on polymer dielectrics. IEEE Transactions on Nuclear Science. 1990;**37**(2):1076-1083. DOI: 10.1109/TNS.1990.574201

[19] Clough R. Nuclear instruments and methods in physics research section B: Beam interactions with materials and atoms. 2001;**185**(1-4):8-33

[20] Chmielewski A, Haji-Saeid M. Radiation technologies: Past, present and future. Radiation Physics and Chemistry. 2004;**71**(1-2):17-21. DOI: 10.1016/j.radphyschem. 2004.05.040

[21] Cleland M, Parks L, Cheng S. Nuclear Instruments and Methods in Physics Research Section B. Applications for radiation processing of materials. 2003;**208**:66-73

[22] Fintzou A, Badeka A, Kontominas M, Riganakos K. Changes in physicochemical and mechanical properties of γ-irradiated polypropylene syringes as a function of irradiation dose. Radiation Physics and Chemistry. 2006;**75**(1):87-97. DOI: 10.1016/j. radphyschem.2005.03.014

[23] Švorčík V, Rybka V, Hnatowicz V, Novotna M, Vognar M. Electron beam modification of polyethylene and polystyrene. Journal of Applied Polymer Science. 1997;**64**(13): 2529-2533. DOI: 10.1002/(SICI) 1097-4628(19970627)64:13< 2529::AID-APP6>3.0.CO;2-F

[24] Hill D, Hopewell J. Effects of 3 MeV proton irradiation on the mechanical properties of polyimide films. Radiation Physics and Chemistry. 1996;**48**(5): 533-537. DOI: 10.1016/0969-806X(96) 00073-4

[25] Mishra R, Tripathy S, Dwivedi K, Khathing D, Ghosh S, Müller M. Fink: Spectroscopic and thermal studies of electron irradiated polyimide. Radiation Measurements. 2003;**36**(1-6):621-624. DOI: 10.1016/S1350-4487(03)00212-9

[26] Sasuga T, Hayakawa N, Yoshida K, Hagiwara M. Degradation in tensile properties of aromatic polymers by electron beam irradiation. Polymer. 1985;**26**(7):1039-1045. DOI: 10.1016/ 0032-3861(85)90226-5

[27] Kucheyev S, Felter T, Anthamatten M, Bradby J. Deformation behavior of ion-irradiated polyimide. Applied Physics Letters. 2004;**85**(5): 733-735. DOI: 10.1063/1.1776618

[28] Mathakari N, Bhoraskar V, Dhole S. MeV energy electron beam induced damage in isotactic polypropylene. Nuclear Instruments and Methods in Physics Research B. 2008;**266**:3075-3080. DOI: 10.1016/j.nimb.2008.03.165

[29] Lappan U, Geißler U, Häußler L, Jehnichen D, Pompe G, Lunkwitz K. Radiation-induced branching and crosslinking of poly (tetrafluoroethylene) (PTFE). Nuclear Instruments and Methods in Physics Research Section B: Beam Interactions with Materials and Atoms. 2001;**185**(1-4):178-183. DOI: 10.1016/S0168-583X(01)00751-0

[30] Millinchuk AV. Sixth international symposium on materials in a space environment. The Netherlands: ESTEC, Noordwijk; 1994. p. 253

[31] Wolan J, Hoflund G. Chemical and structural alterations induced at Kapton® surfaces by air exposures following atomic oxygen or 1 keV Ar+ treatments. Journal of Vacuum Science & Technology A: Vacuum, Surfaces, and Films. 1999;**17**(2):662-664. DOI: 10.1116/1.582027

[32] Milinchuk V, Klinshpont E, Shelukhov I, Smirnova T, Pasevich O.

Degradation of polymer materials in low earth orbits. High Energy Chemistry. 2004;**38**(1):8-12. DOI: 10.1023/B: HIEC.0000012057.54231.9b

[33] Packirisamy S, Schwam D, Litt M. Atomic oxygen resistant coatings for low earth orbit space structures. Journal of Materials Science. 1995;**30**(2): 308-320. DOI: 10.1007/BF00354390

[34] Ferguson D. Interactions between spacecraft and their environments. In: 31st Aerospace Sciences Meeting. USA: Proc. of AIAA; 1993. p. 705. DOI: 10.2514/6.1993-705

[35] de Groh K, Banks B, Hammerstrom A, Youngstrom E, Kaminski C, Marx L, et al. MISSE PEACE polymers: An international space station environmental exposure experiment. USA: Proc. AIAA Conf. on Intl. Space Station Utilizatio; 2001. Available from: http://gltrs.grc.nasa. gov/GLTRS

[36] Zhao X, Shen Z, Xing Y, Ma S. A study of the reaction characteristics and mechanism of Kapton in a plasma-type ground-based atomic oxygen effects simulation facility. Journal of Physics D: Applied Physics. 2001;**34**(15):2308. DOI: 10.1088/0022-3727/34/15/310

[37] Tagawa M, Matsushita M, Umeno M. Proceeding of the Sixth International Symposium on Materials in a Space Environment ESTEC; 19–23 September 1994; Noordwijk, The Netherlands, USA: Proc. of AIAA; pp. 189-193

[38] Milintchouk A, Van Eesbeek M, Levadou F, Harper T. Influence of X-ray solar flare radiation on degradation of teflon-laquo; in space. Journal of Spacecraft and Rockets. 1997;**34**(4): 542-548. DOI: 10.2514/2.3244

[39] Hall D, Fote A. 10 year performance of thermal control coatings at geosynchronous altitude. In: 26th

Thermophysics ConferenceUSA: Proc. of AIAA; 1991. p. 1325. DOI: 10.2514/6.1991-1325

[40] Townsend J, Hansen P, Dever J, de Groh K, Banks B, Wang L, et al. Hubble space telescope metallized teflon (R) FEP thermal control materials: On-orbit degradation and post-retrieval analysis. High Performance Polymers. 1999; **11**(1):81-99. DOI: 10.1088/0954-0083/11/1/007

[41] Skurat VE, Nikiforov AP, Tenovoy AI. Proceedings of the Sixth International Symposium on Materials in a Space Environment ESTS; 19–23 September 1994; Noordwijk, The Netherlands: Proc. of AIAA; pp. 183-187

[42] Majeed R, More S, Phatangare A, Bhoraskar S, Mathe V, Bhoraskar V, et al. Synergetic effects of 1 MeV electron irradiation on the surface erosion in polyimide by atomic oxygen. Nuclear Instruments and Methods in Physics Research Section B: Beam Interactions with Materials and Atoms. 2021;**490**:49-54

[43] Arjun G, Lincy T, Sajitha T, Bhuvaneshwari S, Deepthi T, Devapal D. Atomic oxygen resistant polysiloxane coatings for low earth orbit space structures. Materials Science Forum. 2015;**830**:699-702

[44] Ferguson DC. Interactions between Spacecraft and their Environment. Cleveland, Ohio: NASA Lewis Research Center

[45] de Groh K, Banks B, Hammerstrom A. NASA/TM-2001-211311

[46] Koontz S, Albyn K, Leger L. Atomic oxygen testing with thermal atom systems—A critical evaluation. Journal of Spacecraft and Rockets. 1991;**28**(3): 315-323. DOI: 10.2514/3.26246

[47] Liu T, Sun Q, Meng J, Pan Z, Tang Y. Degradation modeling of

satellite thermal control coatings in a low earth orbit environment. Solar Energy. 2016;**139**:467-474

[48] Reddy M. Effect of low earth orbit atomic oxygen on spacecraft materials. Journal of Materials Science. 1995;**30**(2): 281-307

[49] Ferguson D. The energy dependence and surface morphology of kapton degradation under atomic oxygen bombardment. In: 13th Space Simulation Conference October. USA: Proc. of AIAA; 1984. p. 205

[50] Plis E, Engelhart D, Cooper R, Johnston W, Ferguson D, Hoffmann R. Review of radiation-induced effects in polyimide. Applied Sciences. 2019; **9**(10):1999

[51] Kuroda S, Terauchi K, Nogami K, Mita I. Degradation of aromatic polymers—I. Rates of crosslinking and chain scission during thermal degradation of several soluble aromatic polymers. European Polymer Journal. 1989;**25**(1):1-7. DOI: 10.1016/0014-3057 (89)90200-0

[52] Cherkashina N, Pavlenko V, Abrosimov V, Gavrish V, Trofimov V, Budnik S, et al. Effect of 10 MeV electron irradiation on polyimide composites for space systems. Acta Astronautica. 2021;**184**:59-69. DOI: 10.1016/j.actaastro.2021.03. 032

[53] Alegaonkar PS, Balaya P, Goyal PS, Bhoraskar VN. Dielectric properties of 1-MeV electron-irradiated polyimide. Applied Physics Letters. 2002;**80**:640. DOI: 10.1063/1.1435408

[54] Naddaf M, Balasubramanian C, Alegaonkar P, Bhoraskar V, Mandle A, Ganeshan V, et al. Surface interaction of polyimide with oxygen ECR plasma. Nuclear Instruments and Methods in Physics Research Section B: Beam Interactions with Materials and Atoms.

2004;**222**(1-2):135-144. DOI: 10.1016/j. nimb.2003.12.087

[55] Abdul Majeed R, Purohit V, Bhoraskar S, Mandale A, Bhoraskar V. Irradiation effects of 12 eV oxygen ions on polyimide and fluorinated ethylene propylene. Radiation Effects & Defects in Solids. 2006;**161**(8):495-503. DOI: 10.1080/10420150600810224

[56] Dokhale P, Bhoraskar V, Vijayaraghavan P. A study on boron diffusion in high density polyethylene using the (n, α) reaction. Materials Science and Engineering: B. 1998;**57**(1): 1-8

[57] Grossman E, Gouzman I, Verker R. Debris/micrometeoroid impacts and synergistic effects on spacecraft materials. MRS Bulletin. 2010;**35**(1): 41-47. DOI: 10.1557/mrs2010.615

[58] Abdul Majeed RMA, Datar A, Bhoraskar SV, Alegaonkar PS, Bhoraskar VN. Dielectric constant and surface morphology of the elemental diffused polyimide. Journal of Physics D: Applied Physics. 2006;**39**:4855-4859. DOI: 10.1088/0022-3727/39/22/017

[59] Vieth WR. Diffusion in and Through Polymers: Principles and Applications. Vol. 81. United States: American Scientist; 1991. p. 1

[60] Claeyssens F, Fuge G, Allan N, May P, Ashfold M. Phosphorus carbides: Theory and experiment. Dalton Transactions. 2004;**19**:3085-3092. DOI: 10.1039/B402740J

[61] Zhang R, Wang Z, Yu Y. Analysis of pm-526 inorganic paint by x-ray photoelectron spectroscopy. Chinese Journal of Materials Research. 1990; **4**(2):174-178

[62] Masui T, Hirai H, Hamada R, Imanaka N, Adachi GY, Sakata T, et al. Synthesis and characterization of cerium oxide nanoparticles coated with

turbostratic boron nitride. Journal of Materials Chemistry. 2003;**13**(3): 622-627. DOI: 10.1039/B208109A

[63] Böttcher CJF, Bordewijk P. Theory of Electric Polarization. Vol. 2. Amsterdam, EU: Elsevier Science Limited; p. 1978

[64] Mathakari N, Bhoraskar VN, Dhole SD. 6 MeV pulsed electron beam induced surface and structural changes in polyimide. Materials Science and Engineering B. 2010;**168**:122-126. DOI: 10.1016/j.mseb.2009.11.005

[65] Wilson D, Stenzenzenberger H, Hergenrother P. Polyimides. Glasgow and London: Blackie & Son; 1990

[66] Severin D, Ensinger W, Neumann R, Trautmann C, Walter G, Alig I, et al. Degradation of polyimide under irradiation with swift heavy ions. Nuclear Instruments and Methods in Physics Research Section B: Beam Interactions with Materials and Atoms. 2005;**236**(1-4):456-460. DOI: 10.1016/j.nimb.2005.04.019

[67] Nuclear Instruments and Methods in Physics Research Section B: Beam Interactions with Materials and Atoms. 2005;**236**:1-4, 456-460. DOI: 10.1016/j.nimb.2003.12.051

[68] Cho S, Jun H. Surface hardening of poly (methyl methacrylate) by electron irradiation. Nuclear Instruments and Methods in Physics Research Section B: Beam Interactions with Materials and Atoms. 2005;**237**(3-4):525-532. DOI: 10.1016/j.nimb.2005.03.007

[69] Uelzen T, Müller J. Wettability enhancement by rough surfaces generated by thin film technology. Thin Solid Films. 2003;**434**(1-2):311-315. DOI: 10.1016/S0040-6090(03)00484-X

[70] Vancso G, Hillborg H, Schönherr H. Chemical composition of polymer surfaces imaged by atomic force microscopyand complementary approaches. Polymer Analysis Polymer Theory. 2005;**35**:55-129. DOI: 10.1007/12_046

[71] Flitsch R, Shih D. A study of modified polyimide surfaces as related to adhesion. Journal of Vacuum Science & Technology A: Vacuum, Surfaces, and Films. 1990;**8**(3):2376-2381. DOI: 10.1116/1.576701

[72] Zhang Y, Yang G, Kang E, Neoh K, Huang W, Huan A, et al. Deposition of fluoropolymer films on Si (100) surfaces by Rf magnetron sputtering of poly (tetrafluoroethylene). Langmuir. 2002;**18**(16):6373-6380. DOI: 10.1021/la011606j

[73] Shi M, Selmani A, Martinu L, Sacher E, Wertheimer M, Yelon A. Fluoropolymer surface modification for enhanced evaporated metal adhesion. Journal of Adhesion Science and Technology. 1994;**8**(10):1129-1141. DOI: 10.1163/156856194X00988

[74] Huslage J, Rager T, Schnyder B, Tsukada A. Radiation-grafted membrane/electrode assemblies with improved interface. Electrochimica Acta. 2002;**48**(3):247-254. DOI: 10.1016/S0013-4686(02)00621-7

[75] Ryan M, Fonseca J, Tasker S, Badyal J. Plasma polymerization of sputtered poly (tetrafluoroethylene). The Journal of Physical Chemistry. 1995;**99**(18):7060-7064. DOI: 10.1021/j100018a044

[76] Rasoul F, Hill D, George G, O'Donnell J. A study of a simulated low earth environment on the degradation of FEP polymer. Polymers for Advanced Technologies. 1998;**9**(1):24-30. DOI: 10.1002/(SICI)1099-1581(199801)9:1<24::AID-PAT730>3.0.CO;2-5

[77] Alegaonkar P, Bhoraskar V. Effect of MeV electron irradiation on the free volume of polyimide. Radiation Effects

and Defects in Solids. 2004;**159**(8-9): 511-516

[78] Stelmashuk V, Biederman H, Slavinska D, Zemek J, Trchova M. Plasma polymer films rf sputtered from PTFE under various argon pressures. Vacuum. 2005;**77**(2):131-137. DOI: 10.1021/ma0357164 10.1016/j.vacuum.2004.08.011

[79] Vasilets V, Shandryuk G, Savenkov G, Shatalova A, Bondarenko G, Talroze R, et al. Liquid crystal polymer brush with hydrogen bonds: Structure and orientation behavior. Macromolecules. 2004;**37**(10): 3685-3688. DOI: 10.1021/ma0357164

[80] Williams DH. Fleming. Spectroscopic Methods in Organic Chemistry. 2004

[81] Datar A, Bhoraskar SV, Bhoraskar VN. Surface modification of polymers by atomic oxygen using ECR plasma. Nuclear Instruments and Methods in Physics Research Section B: Beam Interactions with Materials and Atoms. 2007;**258**(2):345-351. DOI: 10.1016/j.nimb.2007.01.230

[82] Sun Y, Zhu Z, Li C. Correlation between the structure modification and conductivity of 3 MeV Si ion-irradiated polyimide. Nuclear Instruments and Methods in Physics Research Section B: Beam Interactions with Materials and Atoms. 2002;**191**(1-4):805-809. DOI: 10.1016/S0168-583X(02)00657-2

[83] Sun Y, Zhang C, Zhu Z, Wang Z, Jin Y, Liu J, et al. The thermal-spike model description of the ion-irradiated polyimide. Nuclear Instruments and Methods in Physics Research Section B: Beam Interactions with Materials and Atoms. 2004;**218**:318-322. DOI: 10.1016/j.nimb.2003.12.051

[84] Vacík J, Červená J, Hnatowicz V, Pošta S, Fink D, Klett R, et al. Simple technique for characterization of ion-modified polymeric foils. Surface and Coatings Technology. 2000;**123**(2-3): 97-100. 10.1016/S0257-8972(99) 00515-0

[85] Stolterfoht N, Fink D, Petrov A, Muller M, Vacik J, Cervena J, et al. Characterization of etched tracks and nanotubules by ion transmission spectrometry. In: Proceedings. 3rd Annual Siberian Russian Workshop on Electron Devices and Materials. Vol. 1. Belarus, Russia: IEEE; 2002. p. 4. DOI: 10.1109/SREDM.2002.1024302

[86] From Solution F. Chemical deposition of chalcogenide thin films from solution. Advances in Electrochemical Science and Engineering. 2008;**12**:165. DOI: 10.1002/9783527616800

[87] Fink D, Alegaonkar PS, Petrov AV, Berdinsky AS, Rao V, Müller M, et al. The emergence of new ion tract applications. Radiation measurements. 2003;**36**(1-6):605-609

[88] Vacik J, Červená J, Hnatowicz V, Fink D, Kobayashi Y, Hirata K, et al. Study of latent and etched tracks by a charged particle transmission technique. Radiation Measurements. 1999;**31**(1-6): 81-84. DOI: 10.1016/S1350-4487(99) 00091-8

[89] Fink D, Klett R. Latent ion tracks in polymers for future use in nanoelectronics: An overview of the present state-of-the-art. Brazilian Journal of Physics. 1995;**25**:1

[90] Herden V. Das Verhalten von lichtinduzierten Ladungsträgern in Polysilane nunterbesonderer Berücksichtigung von Dotierung und strahlenchemischerVernetzung. na. 2001

[91] Krauser J, Weidinger A, Zollondz J, Schultrich B, Hofsäss H, Ronning B. Conducting ion tracks for field emission. In: Proc. Workshop on the

European Network on Ion Track Technology, Caen, France. 2002. pp. 24-26

[92] Lincot D, Froment M, Cachet H. Chemical deposition of chalcogenide thin dlms from solution. Advances in Electrochemical Science and Engineering. 1999;**6**:167-235

[93] Granström M, Berggren M, Inganäs O. Micrometer-and nanometer-sized polymeric light-emitting diodes. Science. 1995;**267**(5203):1479-1481. DOI: 10.1126/science.267.5203.1479

[94] Martin C, Nishizawa M, Jirage K, Kang M, Lee S. Controlling ion-transport selectivity in gold nanotubule membranes. Advanced Materials. 2001; **13**(18):1351-1362. DOI: 10.1002/1521-4095(200109)13:18<1351::AID-ADMA1351>3.0.CO;2-W

[95] Shtanko N, Lequieu W, Du Prez F, Goethals E. Preparation and properties of thermo responsive track membranes. In: Proceedings of the Workshop on European Network on Ion Track Technology. France: Caen; 2002

[96] Desai TA. Nanoporous microfabricated membranes: From diagnostics to drug delivery. In: Proceedings of MRS Fall Meeting. Boston. Germany: Wiley VCH; 27.11.2001–1.12.2001; ContributionY5.5. 2001

[97] Nishizawa M, Mukai K, Kuwabata S, Martin C, Yoneyama H. Template synthesis of polypyrrole-coated spinel LiMn$_2$O$_4$ nanotubules and their properties as cathode active materials for lithium batteries. Journal of the Electrochemical Society;**144**: 1923-1926

[98] Berdinsky A, Fink D, Muller M, Petrov A, Chadderton LT, Apel P Yu. Formation and conductive properties of miniaturized fullerite sensors. Proceedings of MRS Fall Meeting, Boston 27.11.2001–1.12.2001; Contribution Y4.7 2001.

[99] Chakarvarti S, Vetter J. Template synthesis—A membrane based technology for generation of nano-/micro materials: A review. Radiation Measurements. 1998;**29**(2):149-159. DOI: 10.1016/S1350-4487(98)00009-2

[100] Biswas A, Avasthi D, Singh B, Lotha S, Singh J, Fink D, et al. Resonant electron tunneling in single quantum well heterostructure junction of electrodeposited metal semiconductor nanostructures using nuclear track filters. Nuclear Instruments and Methods in Physics Research Section B: Beam Interactions with Materials and Atoms. 1999;**151**(1-4):84-88. DOI: 10.1016/S0168-583X(99)00086-5

[101] Klein J, Herrick R, Palmer D, Sailor M, Brumlik C, Martin C. Electrochemical fabrication of cadmium chalcogenide microdiode arrays. Chemistry of Materials. 1993;**5**:902-904. DOI: 10.1021/cm00031a002

[102] Liu K, Nagodawithana K, Searson P, Chien C. Perpendicular giant magnetoresistance of multilayered Co/Cu nanowires. Physical Review B. 1995; **51**(11):7381. DOI: 10.1103/PhysRevB.51.7381

[103] Hjort K. The European network on ion track technology. In: Proceedings of the Fifth International Symposium on "Swift Heavy Ions in Matter". Sweden: DiVA; May 22–25, 2002; Giordano Naxos, Italy. 2002

Chapter 4

Fluorescent Polyimide in Sensing Applications

Pavitra Rajendran and Erumaipatty Rajagounder Nagarajan

Abstract

Potential advances in sensing can be made by conjugated polymers includes poly(p-phenylene), poly(p-phenylene vinylene), polyfluorene, and poly(thiophene). Among the most important classes of polymers are heterocyclic polymers, such as polyimides, because polyimide nanocomposites possess exceptional mechanical strength as well as chemical, mechanical and temperature resistance. Polyimide offers the potential of providing efficient sensors through its ability to work actively. There is evidence that fluorescent polyimide is efficient at detecting hazardous pollutants. Chemical modifications of the polyimide backbone gave rise to an improved luminescence efficiency of polyimide by incorporating fluorescent chromophores. An overview of recent developments in fluorescent polyimide in sensing applications is presented in this chapter. Some of the fluorescent polyimide materials prepared from different types with surface modification (type-1: perylene tetracarboxylic dianhydride and oxydianiline) (type-2: Tetra (4-aminophenyl) porphyrin and perylenetracarboxylic dianhydride) and (type-3 2-(4,4'-diamino-4''-triphenylamine)-5-(4-dimethylaminophenyl)-1,3,4-oxadiazole) etc. In the following section, the methods and sensing mechanism of fluorescent polyimide are described.

Keywords: Polyimides, composites material, Luminescence, polyimide covalent organic framework

1. Introduction

A chemical sensor converts chemical information into an electrical signal to provide a qualitative or quantitative representation of chemical composition and activity. Chemical sensors create signals by selectively binding a sensing material to an analyte, using a sensing element and a transducer [1]. A sensor's characteristics are dependent upon chemical species that interact with it. Also, it can be either optical or thermal properties of an analyte, which include conductivity, potential, capacity, heat, mass, or optical constant [2]. The simple, convenient and low-cost features of sensor have caused it to receive a great deal of attention [3]. In the medical, biological and environmental fields, sensors have been developed to detect heavy metal ions at low concentrations, an issue of concern for environmental protection and disease prevention. These devices require a high degree of sensitivity and selectivity [4–6]. Optically based sensors are particularly appealing due to their wide range of attractive characteristics, such as microfluidic platforms with integration and the ability to monitor environmental hazards [7, 8]. Recent years have seen a surge in popularity for fluorescent sensors due to their flexibility, quick response times, low detection sensitivity, and simplicity of operation. Optical, electrochemical sensors and biosensors can benefit greatly from conjugated polymers once utilized as sensing materials or chemical probes. There has been an incredible increase in the use of polymer materials such as poly(p-phenylene), poly(p-phenylene vinylene), polyfluorene, and poly(thiophene) in large-area displays, promising future advancements in flexible displays [2, 9–11]. A conjugated polymer backbone or side chain with ionic functional groups produces a conjugated polyelectrolyte that can be combined. They combine the physicochemical properties of polyelectrolytes with those of organic semiconductors, which makes them attractive as materials for sensing, imaging, and device applications [12, 13]. It is possible to synthesize fluorescent polymers by converting fluorescent functional monomers to polymers, using fluorescent compounds as initiators and chain transfer agents, forming chemical bonds between fluorescent groups and polymers, or converting non-fluorescent functional monomers into fluorescent polymers. Fluorescent probes, smart polymers machines, fluorescent chemosensor, fluorescent molecular thermometers, fluorescent imaging, and drug delivery carriers are among their emerging applications in this field. Polymers are advantageous since they can be formed into small particles and thin films that can be coated onto optical fibers, making them ideal for sensors. Several advanced techniques, including electrostatic layer-by-layer assembly and self-assembly of amphiphilic block copolymers with chromophores, have also been employed to manufacture fluorescent systems [14]. The field of thermally stable heteroaromatic polymers is dominated by the polyimides (PI) [15]. Aromatic polyimides are based on aromatic dianhydride and diamine [16]. Among the many different science and engineering fields, that they have been applied in chemistry, physics, electrical and mechanical fields. Aromatic polyimides have received significant attention in recent decades due to their superior physico-chemical, thermal, and mechanical properties [17]. These aromatic polyimide systems are mainly characterized by intramolecular charge transfer (CT) between the diamines and the dianhydrides, and the diamine electrostatic effect dominates their fluorescence properties. As such, it was expected to influence the intensity and strength of charge transfer and donor-acceptor interactions along the polymer backbone [11, 18]. Similarly, a promising type of porous crystalline material known as covalent organic frameworks (COFs) has gained popularity and have displayed promise in various applications, including gas storage, chemosensor, catalysis, and electricity. Meanwhile, COF combined with polyimide (PI) shows great thermal stability, good crystalline structure, large pore sizes, and high surface areas. While synthesis of various PI-COF units with unique properties is still an unsolved problem, there has been no report on PI-COFs that possess fluorescent features, or their application to chemical and biological sensing systems. A novel covalent organic framework of fluorescent polyimide which has been shown to provide enhanced detection of 2,4,6-trinitrophenol (TNP) with excellent sensitivity and selectivity [19]. A method allowing selective chemosensitive Fe^{3+} detection with the COFs has been developed [20]. Recent research has focused mainly on developing and applying polyimide-based fluorescent sensors for food detection. Likewise, polyimide nanocomposite, piezoelectric sensors, and in particular, surface acoustic wave (SAW) biosensors are gaining attention due to their rapid response and non-labeling capabilities for detecting macromolecules in biological systems. Though the mechanism of fluorescent polyimide is not fully understood, it's believed that charge transfer effect is an important factor. Thus, few researchers have focused on increasing the quantum efficiency of aromatic polyimides by regulating the charge transfer CT effect.

2. Fluorescent polyimides

Polyimides is widely used in the field of sensor, primarily humidity sensor [21], strain sensor [22], gas sensor [23] and bio sensor [24]. Recently aromatic polyimide have gotten incredible consideration because of their predominant physical, chemical, thermal and mechanical properties like resistance to chemical and radiation damages, mechanical durability, temperature tolerance, and mobility. Aromatic polyimides show high-performance engineering polymers, accordingly prompts the use across various science and engineering disciplines. The strong superiority of aromatic polyimides over conventional conjugated polymers, however, has led to a relatively low utilization of these materials in light-emitting devices, as they are typically non-luminous traditional polyimides or luminance with low luminescence efficiency. As a means to enhance the intensity of luminescence of aromatic polyimides. In polyimide backbones or as pendant groups, fluorescent chromophores have been integrated in order to optimize fluorescence properties [11, 17, 25]. Fluorescence quenching, however, occurred because the diamine and dianhydride moieties were undergoing strong intermolecular and intramolecular charge transfer between their occupied and unoccupied molecular orbitals, and there were strong π-π interactions between chromophores. Fluorescence quenching was resolved by using aliphatic monomers. The fluorescence intensity was boosted by hindering the formation of charge-transfer-complexes (CTC) in non-conjugated and low electron-donating groups. Hence, fluorescence has been acquired by the addition of triphenylamine moieties into the polyimide backbone [11]. Also, new kind of polymer developed by the introduction of electroluminescent active organic dye unit into the polyimide backbone. For example, 2,5-distyrylpyrazine, electroluminescent active organic dye unit was incorporated into the polyimide backbone [26]. Polymers made in this way are typically synthesized by reacting dianhydride with a diamine under optimal conditions. To extend the use of polyimide, novel polyimide with different chemical moieties were reported in literature. Like, a novel diphenyl-fluorene-based Cardo copolyimide containing perylene, synthesized by polycondensation of a diamine 4,4'-(9H-fluoren-9-ylidene)diphenylamine with perylene dianhydride and dianhydride in *m*-cresol with isoquinoline as catalyst at 200°C [27]. A novel aromatic polyimides containing 4,5-diazafluorene were synthesized using dihydride monomer, 9,9-di[4-(3,4-dicarboxyphenoxy)phenyl]-4,5-diazafluorene dianhydride [28]. A fluorescent polyimide was formulated by reacting perylene tetracarboxylic dianhydride and oxydianiline in N-methyl pyrrolidone solvent under ideal conditions [17]. Developing high-performance flexible light-emitting materials would be greatly enhanced by studies on these materials. As is widely known, polyimides have great optical properties due to their charge-transfer (CT) behavior, especially rigid polyimides with lots of aromatic compounds. Polyimide with high fluorescence efficiency by control of the push-pull relationship *via* electrostatic interactions between the diamine moieties and the dianhydride moieties [11] developed two different polyimide, they were pyrrole-containing polyimide (PyODPI) and aromatic polyimide containing the triazole group (TzODPI) using two different diamine monomers 4,40 - (1-(4-tritylphenyl)-pyrrole-2,5-diyl)dianiline and 4,40 - (4-(4-tritylphenyl)-1,2,4-triazole-3,5-diyl)dianiline with identical chemical structures that differ in their electronic effects respectively. This aromatic polyimide contains a triazole moiety, as shown by its bright green photoluminescence has a quantum yield as high as 61%, but as a result of the film formation, it has a quantum yield of 13%. The pyrrole-containing polyimide PyODPI, however, completely slowed down the fluorescence. Polymers displayed a completely different fluorescence behavior, since 4,40-(1-(4-tritylphenyl)-pyrrole-2,5-diyl)dianiline monomer contains electron-donating pyrrole groups while 4,40-(4-(4-tritylphenyl)-1,2,4-tri-azole-3,5-diyl)dianiline monomer carries an electron acceptor 1,2,4-triazole group. A major influence on fluorescence properties was the electronic effect of diamines, which may have controlled the strength of charge transfer and donor-acceptor interaction intensity at polymer backbones. The mechanism for fluorescence has been clarified by computing the orbital distribution and oscillator strength, as well as electron transitions between the ground and excited states of a model molecule. The findings show that a deficiency of electrons in diamine groups improves photoluminescence efficiency by inhibiting the charge-transfer processes that cause fluorescence to be diminished [11].

2,4,6-Trinitrophenol (TNP) is highly explosive and one of the most dangerous nitro-aromatic explosives and is superior to its more conventional counterpart 2,4,6-Trinitrotoluene (TNT). The solvothermal synthesis of a fluorogenic polyimide covalent organic framework (PI-COF) for the detection of 2,4,6-trinitrophenol has been accomplished using tetra(4-aminophenyl) porphyrin and perylenetracarbox-ylic dianhydride. A key quality of PI-COF is its porous crystalline structure and excellent thermal stability (above 500°C). In addition to providing highly sensitive and selective detection, PI-COFs can be utilized as efficient fluorescent probes for the detection of 2,4,6-trinitrophenol (TNP). The phenomenon might be caused by the combination of electron transfer and inner-filter effects, according to DFT calculations and spectral overlap data. As TNP is detected, a linear response between 0.5 to 10 µM is observed with a detection limit of 0.25 µM [19].

Two new polyimide-based porous covalent organic frameworks has been developed and studied their properties [20]. Those compounds were synthesized by directly heating mixtures of melamine and pyromellitic dianhydride, as well as naphthalenetetracarboxylic dianhydrides in N_2 atmosphere, respectively. The strong fluorescence of these two PI-COFs was due to their high electro-delocalization and inherent rigidity of COF, which resulted from the fluorescence transition $\pi^* \rightarrow n$ induced by the proper solvents. In response to the strong quenching effects of Fe^{3+} on the fluorescence of the COFs, Fe^{3+}-specific chemosensing was achieved. The $n - \pi^*$ transition in N,N-dimethylformamide and the alkaline aqueous solution caused strong fluorescence to be observed in both COF-1 and COF-2 of π-conjugated frameworks. In real samples, a high linear correlation coefficient, a wide linear range, and a low detection limit were achieved. Aggregation effect and π–π reaction could be responsible for fluorescence quenching [20].

An innovative nanofibrous membrane composed of porphyrinated polyimide (PPI) that can detect trace amounts of hydrogen chloride (HCl) gas rapidly. Porphyrin fluorophores can be incorporated covalently into the main chains of polyimide to overcome the porphyrin aggregation and enhance polyimide's physicochemical stability. The dual chromogenic and fluorogenic properties of the nanofibrous membrane in the presence of HCl gas influence its optical properties through distortions created by the out-of-plane behavior of the macrocyclic porphyrin. Based on calorimetric and fluorimetric analysis, it is readily apparent that the color changes evident are in response to HCl gas. A calculated affinity constant indicates that the nanofibrous membrane sensor has a distinct sensitivity to the presence of HCl, as determined by SPR analysis $(1.05 \pm 0.23) \times 10^4$ L mol^{-1}. The PPI nanofiber membrane sensor is found to be highly sensitive to HCl, based on the apparent binding affinity constant of $1.05 \pm 0.23 \times 10^4$ L mol^{-1}. A nanofibrous membrane sensor designed using PPI also has improved thermal stability, making it attractive for monitoring emissions of HCl gas from incinerators that burn household, clinical or industrial waste [29].

A large-area and highly porous nanofibrous polymer membrane possesses high sensitivity, and a rapid response time is a key aspect of sensing applications. The study [30] presents a novel zinc porphyrin-containing polyimide (ZPCPI) nanofibrous membrane capable of detecting trace amounts of pyridine vapor rapidly and reversibly. The analyte can be detected at concentrations as low as 0.041 ppm. It is due to the high chemical and thermal resistance of polyimide, as well as its mechanical stability, that pyridine vapor can be detected in harsh environments with zinc porphyrin fluorophore. ZP's photophysical responses can be attributed to changes in its electronic state as well as geometric distortion caused by strong pyridine coordination. Pyridine possesses both high gas-phase basicity and small molecular size, thus perfectly suited for sensors that select pyridine vapor in comparison with different amines and other gases that could interfere [30].

A large number of porphyrinated nanofibers could be used as sensing materials because of porphyrin's large versatility in polymer synthesis. A porphyrinated polyimide nanofiber material with unique luminescent properties turned out to be an effective sensory material for detecting TNT vapor (10 ppb) in its trace form. Polyimide nanofibers are improved in terms of physicochemical stability when porphyrin fluorophores are covalently bonded to the main chains: this reduces the aggregation-caused self-quenching of porphyrin fluorescence. Porphyrinated nanofibers have a large surface area-to-volume ratio and excellent gas accessibility, which result in a much more impressive fluorescent quenching behavior towards trace TNT than their spin-coated dense film counterparts. The quenching efficiency of the other compounds, such as 2,4-dinitrotoluene (DNT), 2,4,6-trinitrophenol (Picric acid: PA) and nitrobenzene (NB) is much lower than that of TNT. From SPR analysis, porphyrinated nanofibers have an apparent affinity constant of $(2.37 \pm 0.19) \times 10^7$ L/mol, which implies that they are a promising alternative for TNT detection [31].

The synthetic method yields 2-(4,4'-diamino-4''-triphenylamine)-5-(4-dimethylaminophenyl)-1,3,4-oxadiazole, a novel blue fluorescent aromatic diamine. A polyimide and three poly(amide-imide)s consisting of the fluorescent imide-type polymers are produced using polycondensation reactions based on diamine. Aromatic polyimides with amide groups in their polymer chains are classified as aromatic poly(amide-imides), which exhibit higher thermal stability, greater solubility, and lower glass transition temperatures than aromatic polyimides of similar chemical structure. The 1,3,4-oxadiazole rings present in the polymers' macromolecular chains make them more thermo-oxidatively stable, more hydrolytically resistant, and mechanically strong. As 1,3,4-oxadiazole rings have electron-withdrawing properties, several polymers containing oxadiazole were investigated for use in the production of organic light emitting diodes. Light-emitting diodes are likely to utilize materials with electron-donor modified dimethylamino substituents in *para*-positions of the pendant chromophoric 2,5-diphenyl-1,3,4-oxadiazole unit. Due to their ability to transport electrons and holes, these substituted diphenyl-1,3,4-oxadiazoles display intense fluorescence. Fluorescence was observed in the blue region with a high quantum yield and a large Stokes shift value. With HCl as a dopant, protonation led to a noticeable decrease in fluorescent intensity, caused by nitrogen atoms that have free electron pairs, derived from 1,3,4-oxadiazole rings and dimethylamino groups [32].

Calorimetric devices can be produced simply, accurately, and eco-friendly by directly using film-based polyimide sensors. Although few studies have been published on film-based PI sensors capable of detecting water in organic solvents *via* calorimetry. The authors [33] present a synthetic pathway for synthesis of two PIs containing hydroxyl groups that exhibit excellent synergistic effects. The fluoride ion-induced deprotonated species of hydroxyl-containing polyimide can quickly be reprotonated in the presence of water, making it an ideal water-sensing material. In the similar way, deprotonated fluoride ion-induced PIs that contain hydroxyls are easily protonated when exposed to water, which was used as a water sensor [33].

Hybridized local and charge transfer transitions produce white light intrinsically in semi-aliphatic hyperbranched polyimides with epoxide terminal groups (EHBPI). Studies of solution fluorescence indicate the following: 1. Fluorescence is influenced by the backbone structure and the terminal chemical groups at the end of the polymer, 2. Concentration and temperature are key factors influencing solution fluorescence and 3. Copper and iron ions quench solution fluorescence. As well as the increasing drop in quantum yield with higher concentrations or the addition of water, and higher quantum yield found in the solution-state as compared to the solid-state. The semi-aliphatic hyperbranched polyimides, which possess epoxide terminal groups, emits bluish green fluorescence when in solution, and white fluorescence when it is solid. Additionally, a white fluorescence can be detected when the free-standing film is prepared by blending EHBPI with poly(vinyl alcohol). Despite its simplicity, single-component hyperbranched polymers could offer unique benefits in certain applications since they are solid-state white fluorescence. In the UV/Vis absorption spectra and DFT calculations, the hybridized local and charge transfer (HLCT) transition is found to be the lowest electronic transition for semi-aliphatic hyperbranched polyimides, whereas the charge transfer transition is found for aromatic hyperbranched polyimides. Upon excitation, semi-aliphatic hyperbranched polyimides exhibit visible fluorescence due to the hybridized local and charge transfer (HLCT) transition. HBPIs fluoresced more intensely when their terminal phenolic groups were converted to epoxide groups. An intrinsic white-light-emitting hyperbranched polyimide could therefore be used to sense 'turn-off' sensors for iron (III). Polymer architecture appears to have complex effects on fluorescence, raising the need for a more targeted molecular design to fully exploit the hyperbranched architecture [34].

3. Polyimide biosensor

Recent research has shown that conducting polymers incorporating nanomaterials can be used as transducing media in biosensing systems [35]. A high surface area to volume ratio of conducting polymer nanocomposites becomes very sensitive and they show outstanding performance combining them with biomarkers yields remarkable selectivity [36]. In addition to the ability of nanoparticles to act as electron connectors, while matrix polymers can also assist with the adsorption of a target substance. The thin film of conducting polymer nanocomposite thus becomes an effective transducing electrode for biosensors [37]. Through electrochemical entrapment, covalent immobilization, or affinity interactions, a thin film of conducting polymers can serve as a medium for immobilizing biomolecules [38]. The immobilized biomolecules bind to analytes through the adsorption process. Those effects may manifest themselves in change in mass, optical activity, electrical conductivity, and temperature around electrode surfaces. This type of change can be detected with the appropriate detector. Piezoelectric biosensors, in particular, surface acoustic wave (SAW) biosensors, are widely used today for sensing biological macromolecules due to their real-time, label-free, and highly sensitive detection capabilities [39]. Also, SAW devices have gained popularity because of their low cost, compact size, and ease of analysis. Especially when it comes to sensitivity, selectivity, and stability, SAW biosensors are greatly impacted by the performance of the bioreceptor surfaces. The main disadvantages of polymer-based SAW devices are their high insertion losses and, therefore, low analytical window. An effective solution can be provided by a conducting polymer nanocomposite system with nanoparticles that act in synergy with the matrix polymer to conduct/transduce acoustic waves. First time, polyimide nanocomposite (PI/AuNP-MoS_2-rGO) was studied as a bioreceptor base used for Carcinoembryonic antigen (CEA) detection. CEA is a tumor-marking protein that is found in several types of cancer. In addition, the thin film of polymer nanocomposite was turned into a bioreceptor by immobilizing CEA antibodies (anti-CEA) through thioglycolic acid bridges and activation with EDC-NHS. Through covalent bonding between the Mercapto part and Au, AuNP can serve as a host for the thioglycolic acid bridge groups. Polyimide nanocomposite showed reliability and stability of the device over time. The biosensor was found to have a limit of detection (LOD) of 0.084 ng/mL. A validation study validated the real-time capabilities of the biosensor by analyzing clinical serum samples and analyzing its selectivity by demonstrating its affinity for other common cancer-marking proteins. Likewise, the biosensor displayed excellent stability, with only 10% reduction in activity recorded until the 80th day of storage [40]. Paraoxon detection was achieved by synthesizing an acetylcholinesterase (AChE) film biosensor based on reduced graphene oxide/polyimide thin films (rGO/PI). By using a modified AuNPs-MoS$_2$-rGO/PI flexible film biosensor, acetylcholine chloride was hydrolyzed successfully to obtain a large current response at 0.49 V and demonstrate successful immobilization of AChE. The AChE/AuNPs-MoS$_2$-rGO/PI film biosensor displays a linear response over a concentration range of 0.005–0.150 µg/mL, 4.44 uA/µg/ mL of sensitivity, 0.0014 µg/mL of detection limit, good reproducibility and stability on the paraoxon inhibition on the AChE [41]. There are a number of applications for flexible biosensors in measuring the concentration of target bio-analytes. Additionally, to their flexibility, electrochemical sensors made with 2D materials also have several advantages such as scalability, increased compatibility across a wider area, and greater transparency. Molybdenum disulfide (MoS$_2$) on a polyimide (PI) substrate was used to fabricate a flexible biosensor that can be used in electrochemistry platforms. Because MoS$_2$ has a higher electrical conductivity, a flexible MoS$_2$–Au–PI sensor can provide highly sensitive detection of target proteins with a relatively quick response via cyclic voltammetry. This device can detect hormones such as triiodothyronine (T3), endocrine-related hormones parathyroid hormone (PTH), and thyroxine (T4) with a high degree of sensitivity, as well as locate their position with a high degree of accuracy [24]. Composite membrane sensors made with molybdenum doped reduced graphene oxide and polyimide (Mo-rGO/PI) exhibit good catalytic activity for dopamine (DA), with linear responses from the range of 0.1 to 2000 µM. Furthermore, the assay has good stability and reproducibility, and it is effective at detecting DA in human blood serums [42]. An Au nanoparticle decorated polyimide electrochemical sensor for uric acid has been developed using its precursors, 4,4′(4.4′-isopropylidene-diphenoxy) bis (phthalic anhydride) and aniline tetramer. A nanoparticle of gold was then incorporated into electroactive polyimide. This sensor has the best sensitivity of 1.53 µM, a detection limit of 0.78 µM, and a linear measurement range is about 5 to 50 µM at 310 mV. Au nanoparticles decorated polyimide exhibited the best selectivity for UA, dopamine (DA), and ascorbic acid (AA) [43]. Known for their high surface area, carbon nanotube (CNTs) is also known to have excellent mechanical properties, good electrical properties, and good thermal conductivity. Composite materials with CNT and polymer offer desirable mechanical and electrical properties because CNT can enhance properties of composites. The PI/CNT membrane is superior to the conventional electrodes like indium tin oxide and glassy carbon electrode (GCE) due to its superior conductivity and mechanical properties. GCE is commonly used to construct sensors, however, it has low conductivity. Nanomaterials like Ni(OH)$_2$ are directly deposited on PI/CNT membranes, and Ni(OH)$_2$ and PI/CNT membrane work as a sensor without any substrate electrodes. Because of its good conductivity, PI/CNT membranes, which improves Ni(OH)$_2$ sensing, as well as the electron transfer, therefore

PI/CNT-Ni(OH)$_2$ sensors are most suitable for detecting glucose. It has a number of notable qualities, including a good stability, high selectivity and sensitivity, and rapid amperometric response. PI/CNT–Ni(OH)$_2$ exhibit 2071.5 μA mM^{-1} cm^{-2} of high sensitivity and 0.36 mM of detection limit at +0.60 V [44]. Ni(OH)$_2$/MoS$_x$ / CNT/PI sensors advantages included a wide linear range from10 to 1600 μM glucose, quick response, a minimum detection limit of 5.4 μM, high selectivity, reliable reproducibility, and long-term durability up to two weeks. Ni(OH)$_2$ and MoS$_x$ have a pronounced synergistic effect that explains the superior performance. Ni(OH)$_2$/ MoS$_x$ /CNT/PI sensors for measuring blood glucose [45]. A polyimide (PI)-boron nitride (BN) composite is a dopamine-selective membrane. A polyimide matrix with BN particles showed better porosity, selectivity, and thermal resistance than polyimides without BN particles. A high degree of sensitivity, reversibility, and a low detection threshold (4 × 10^{-8} M) were all characteristics of %BN. Polyimide membranes have extremely high R values (0.9904) [46].

4. Conclusions

Its superior physical, chemical, thermal and mechanical properties have made aromatic polyimide a popular material across a wide range of industries. Though there is no complete understanding of the mechanism of fluorescent polyimide, it is thought that charge transfer action plays a crucial role. As a result of intramolecular charge transfer (CT) between diamines and dihydrides, aromatic polyimide systems have strong fluorescence properties, and their properties are largely governed by the diamine electrostatic effect. Fluorescence of aromatic polyimide depends on the intensity, strength, and donor-acceptor interactions along the polymer backbone. Until recently, few researchers have attempted to increase the quantum efficiency of aromatic polyimides by regulating the CT effect. Development of an efficient food detection system using polyimide-based fluorescent sensors is being driven primarily by experimental research. A broad range of polyimide nanocomposite, piezoelectric and surface acoustic wave (SAW) biosensors have gained attention due to their high sensitivities, real-time and label-free functionality.

Acknowledgements

The author, P.R., acknowledges Kalasalingam Academy of Research and Education for providing research fellowship and necessary facilities.

Conflict of interest

The authors declare that they have no conflict of interest.

Author details

Pavitra Rajendran and Erumaipatty Rajagounder Nagarajan*
Department of Chemistry, Kalasalingam Academy of Research and Education,
Krishnankoil, Tamil Nadu, India

*Address all correspondence to: nagarajanklu@gmail.com

IntechOpen

References

[1] Faridbod F, Ganjali M R, Hosseini M. 12 - Lanthanide materials as chemosensors. Lanthanide-Based Multifunctional Materials. Martín-Ramos P and Ramos Silva M, Eds. Elsevier; 2018, p. 411-454. DOI: 10.1016/B978-0-12-813840-3.00012-0

[2] Abdul S, Judit T, Ilona F, Nikoletta M. Chapter 16 - Functional thin films and nanostructures for sensors. Fundamentals of Nanoparticles. Barhoum A, Hamdy Makhlouf A S, Eds. Elsevier; 2018, p. 485-519. DOI: 10.1016/B978-0-323-51255-8.00016-1

[3] Zhou Z, Niu W, Lin Z, Cui Y, Tang X, Li Y. A novel 'turn-off' fluorescent sensor for Al3+ detection based on quinoline carboxamide-coumarin. Inorganic Chemistry Communications. 2020; 121,108168. DOI: 10.1016/j.inoche.2020.108168

[4] Farzin L, Shamsipur M, Sheibani S. A review: Aptamer-based analytical strategies using the nanomaterials for environmental and human monitoring of toxic heavy metals. Talanta. 2017; 174. 619-627. DOI: 10.1016/j.talanta.2017.06.066.

[5] Nolan E M, Lippard S J. A 'Turn-On' Fluorescent Sensor for the Selective Detection of Mercuric Ion in Aqueous Media. J. Am. Chem. Soc. 2003; 125,14270-14271, DOI: 10.1021/ja037995g

[6] Ye B C, Yin B C, Highly Sensitive Detection of Mercury(II) Ions by Fluorescence Polarization Enhanced by Gold Nanoparticles. Angewandte Chemie International Edition. 2008; 47. p. 8386-8389. DOI: 10.1002/anie.200803069

[7] Aiestaran P, Dominguez V, Arrue J, Zubia J, A fluorescent linear optical fiber position sensor. Optical Materials. 2009; 31. p. 1101-1104. DOI: 10.1016/j.optmat.2007.12.022

[8] Kuswandi B, Nuriman, Huskens J, Verboom W, Optical sensing systems for microfluidic devices: A review. Analytica Chimica Acta. 2007; 601. p. 141-155. DOI: 10.1016/j.aca.2007.08.046

[9] Tan D, He Y, Xing X, Zhao Y, Tang H, Pang D, Aptamer functionalized gold nanoparticles based fluorescent probe for the detection of mercury (II) ion in aqueous solution. Talanta. 2013;113. p. 26-30. DOI: 10.1016/j.talanta.2013.03.055

[10] Vendrell M, Zhai D, Er J C, Chang Y T, Combinatorial Strategies in Fluorescent Probe Development. Chem. Rev. 2012; 112. p. 4391-4420, DOI: 10.1021/cr200355j

[11] Zhou Z, Zhang Y, Liu S, Chi Z, Chen X, Xu J, Flexible and highly fluorescent aromatic polyimide: design, synthesis, properties, and mechanism. J. Mater. Chem. C. 2016; 4. p. 10509-10517. DOI: 10.1039/C6TC03889A

[12] Chen A. Wu W, Fegley M, Pinnock S, Duffy-Matzner J, Bernier W, Jones W, Pentiptycene-Derived Fluorescence Turn-Off Polymer Chemosensor for Copper(II) Cation with High Selectivity and Sensitivity. Polymers. 2017; 9. 118. DOI: 10.3390/polym9040118

[13] Liu B, Tang B Z, Fluorescent Sensors, Macromolecular Rapid Communications. 2013; 34. p. 704-704, 2013, DOI:10.1002/marc.201300077

[14] Yuan J, Wang S, Shan J, Peng J, Wei L, Xu X, Formation and Photoluminescence of Fluorescent Polymers. International Journal of Polymer Science. 2010; 2010. p. e526348. DOI: 10.1155/2010/526348.

[15] Wright W W, Hallden-Abberton M, Polyimides. Ullmann's Encyclopedia of Industrial Chemistry. Wiley-VCH Verlag

GmbH & Co. KGaA, Ed. Weinheim, Germany: Wiley-VCH Verlag GmbH & Co. KGaA. 2000; p. a21_253. DOI: 10.1002/14356007.a21_253.

[16] "Polyimide (PI) Plastic: Uses, Structure, Properties & Applications." https://omnexus.specialchem.com/ selection-guide/polyimide-pi-plastic

[17] Sribala G, Meenarathi B, Anbarasan R, Synthesis, characterization, and catalytic activity of fluorescent polyimide nano composites. J. Appl. Polym. Sci. 2017; 134.12. DOI:10.1002/app.44633

[18] Li Q, Horie K, Yokota R, Absorption, Fluorescence, and Thermal Properties of Transparent Polyimides Based on Cyclobutane tetracarboxylic Dianhydride. Polym J. 1998; 30. p. 805-812. DOI: 10.1295/polymj.30.805

[19] Zhang C, Zhang S, Yan Y, Xia F, Huang A, Xian Y, Highly Fluorescent Polyimide Covalent Organic Nanosheets as Sensing Probes for the Detection of 2,4,6-Trinitrophenol. ACS Appl. Mater. Interfaces. 2017; 9. p. 13415-13421. DOI: 10.1021/acsami.6b16423

[20] Wang T, Xue R, Chen H, Shi P, Lei X, Wei Y, Guo H, Yang W, Preparation of Two New Polyimide Bond Linked Porous Covalent Organic Frameworks and Their Fluorescence Sensing Application for Sensitive and Selective Determination of Fe^{3+}. New J. Chem. 2017; 41. p.14272-14278. DOI:10.1039/C7NJ02134H

[21] Boudaden J, Steinmaßl M, Endres H E, Drost A, Eisele I, Kutter C, Müller-Buschbaum P, Polyimide-Based Capacitive Humidity Sensor. Sensors. 2018. 18.1516. DOI:10.3390/s18051516

[22] Jiang Y, He Q, Cai J, Shen D, Hu X, Zhang D, Flexible Strain Sensor with Tunable Sensitivity via Microscale Electrical Breakdown in Graphene/ Polyimide Thin Films. ACS Appl. Mater. Interfaces. 2020;12. p. 58317-58325. DOI: 10.1021/acsami.0c19484.

[23] Padua L M G, Yeh J M, Santiago K S, A Novel Application of Electroactive Polyimide Doped with Gold Nanoparticles: As a Chemiresistor Sensor for Hydrogen Sulfide Gas. Polymers. 2019.11. DOI: 10.3390/ polym11121918

[24] Kim H U, Kim H Y, Seok H, Kanade V, Yoo H, Park K Y, Lee J H, Lee M H, Kim T, Flexible MoS2–Polyimide Electrode for Electrochemical Biosensors and Their Applications for the Highly Sensitive Quantification of Endocrine Hormones: PTH, T3, and T4. Anal. Chem. 2020; 92. 6327-6333. DOI: 10.1021/acs.analchem.9b05172

[25] Liu Q, Paul D R, Freeman B D, Gas permeation and mechanical properties of thermally rearranged (TR) copolyimides. Polymer. 2016; 82. p. 378-391. DOI: 10.1016/j.polymer.2015. 11.051

[26] Wu A, Akagi T, Jikei M, Kakimoto M, Imai Y, Ukishima S, Takahashi Y, New Fluorescent Polyimides for Electroluminescent Devices Based on 2,5-Distyrylpyrazine. *Thin Solid Films.* 1996; 273. 214-217. DOI:10.1016/0040-6090(95)06780-9

[27] Yang M, Xu S, Wang J, Ye H, Liu X, Synthesis, characterization, and electroluminescent properties of a novel perylene-containing copolyimide. Journal of Applied Polymer Science. 2003; 90. p. 786-791. DOI: 10.1002/ app.12527

[28] Deng B, Zhang S, Liu C, Li W, Zhang X, Wei H, Gong C, Synthesis and Properties of Soluble Aromatic Polyimides from Novel 4,5-Diazafluorene-Containing Dianhydride. RSC Adv. 2017; 8. 194-205. DOI:10.1039/C7RA12101F

[29] Lv Y Y, Wu J, Xu Z K, Colorimetric and fluorescent sensor constructing

from the nanofibrous membrane of porphyrinated polyimide for the detection of hydrogen chloride gas. Sensors and Actuators B: Chemical. 2010.148. p. 233-239. DOI: 10.1016/j. snb.2010.05.029

[30] Y. Lv, Y. Zhang, Y. Du, J. Xu, and J. Wang, "A Novel Porphyrin-Containing Polyimide Nanofibrous Membrane for Colorimetric and Fluorometric Detection of Pyridine Vapor," Sensors, vol. 13, no. 11, Art. no. 11, Nov. 2013, doi: 10.3390/s131115758.

[31] Lv Y Y, Xu W, Lin F W, Wu J, Xu Z K, Electrospun nanofibers of porphyrinated polyimide for the ultra-sensitive detection of trace TNT. Sensors and Actuators B: Chemical. 2013; p. 205-211. DOI: 10.1016/j. snb.2013.04.094.

[32] Hamciuc C, Hamciuc E, Homocianu M, Nicolescu A, Lisa G, New blue fluorescent and highly thermostable polyimide and poly(amide-imide)s containing triphenylamine units and (4-dimethylaminophenyl)-1,3,4-oxadiazole side groups. Dyes and Pigments. 2018; 148. p. 249-262. DOI: 10.1016/j.dyepig.2017.09.010

[33] Wu Y, Ji J, Zhou Y, Chen Z, Liu S, Zhao J, Ratiometric and colorimetric sensors for highly sensitive detection of water in organic solvents based on hydroxyl-containing polyimide-fluoride complexes. Analytica Chimica Acta. 2020; 1108. p. 37-45. DOI: 10.1016/j. aca.2020.02.043

[34] Xing A, Miao X, Liu T, Yang H, Meng Y, Li X, An intrinsic white-light-emitting hyperbranched polyimide: synthesis, structure–property and its application as a 'turn-off' sensor for iron(iii) ions, J. Mater. Chem. C. 2019; 7. p. 14320-14333. DOI: 10.1039/C9TC04102H

[35] Shrivastava S, Jadon N, Jain R, Next-generation polymer nanocomposite-based electrochemical sensors and biosensors: A review. TrAC Trends in Analytical Chemistry. 2016; 82. p. 55-67. DOI: 10.1016/j.trac.2016.04.005

[36] Lu L, Zhu Z, Hu X, Hybrid nanocomposites modified on sensors and biosensors for the analysis of food functionality and safety. Trends in Food Science & Technology. 2019; 90.p. 100-110. DOI: 10.1016/j.tifs.2019.06.009

[37] Zhang P, Sun T, Rong S, Zeng D, Yu H, Zhang Z, Chang D, Pan H, A Sensitive Amperometric AChE-Biosensor for Organophosphate Pesticides Detection Based on Conjugated Polymer and Ag-RGO-NH2 Nanocomposite. Bioelectrochemistry; 2019; 127. 163-170. DOI: 10.1016/j. bioelechem.2019.02.003.

[38] Azharudeen A M, Karthiga R, Rajarajan M, Suganthi A, Fabrication, characterization of polyaniline intercalated NiO nanocomposites and application in the development of non-enzymatic glucose biosensor. Arabian Journal of Chemistry. 2020; 13. p. 4053-4064, DOI: 10.1016/j.arabjc.2019.06.005.

[39] Skládal P, Piezoelectric biosensors, TrAC Trends in Analytical Chemistry. 2016; 79. p. 127-133. DOI: 10.1016/j.trac.2015.12.009.

[40] Jandas P J, Luo J, Prabakaran K, Chen F, Fu Y Q, Highly stable, love-mode surface acoustic wave biosensor using Au nanoparticle-MoS2-rGO nano-cluster doped polyimide nanocomposite for the selective detection of carcinoembryonic antigen. Materials Chemistry and Physics. 2020; 246. p. 122800. DOI: 10.1016/j.matchemphys.2020.122800

[41] Jia L, Zhou Y, Wu K, Feng Q, Wang C, He P, Acetylcholinesterase modified AuNPs-MoS2-rGO/PI flexible film biosensor: Towards efficient

fabrication and application in paraoxon detection. Bioelectrochemistry. 2019; 131. p. 107392. DOI: 10.1016/j. bioelechem.2019.107392

[42] Jia L, Zhou Y, Jiang Y, Zhang A, Li X, Wang C, A novel dopamine sensor based on Mo doped reduced graphene oxide/polyimide composite membrane. Journal of Alloys and Compounds. 2016; 685. p. 167-174. DOI: 10.1016/j. jallcom.2016.05.239

[43] Bibi A, Hsu S C, Ji W F, Cho Y C, Santiago K S, Yeh J M, Comparative Studies of CPEs Modified with Distinctive Metal Nanoparticle-Decorated Electroactive Polyimide for the Detection of UA. Polymers. 2021; 13. DOI: 10.3390/polym13020252

[44] Jiang Y, Yu S, Li J, Jia L, Wangs C, Improvement of sensitive Ni(OH)2 nonenzymatic glucose sensor based on carbon nanotube/polyimide membrane. Carbon. 2013; 63. p. 367-375. DOI: 10.1016/j.carbon.2013.06.092

[45] Wang Q, Zhang Y, Ye W, Wang C, Ni(OH)2/MoSxnanocomposite electrodeposited on a flexible CNT/PI membrane as an electrochemical glucose sensor: the synergistic effect of Ni(OH)2 and MoSx. J Solid State Electrochem. 2016; 20. p. 133-142. DOI: 10.1007/s10008-015-3002-9

[46] Aksoy B, Paşahan A, Güngör Ö, Köytepe S, Seçkin T, A novel electrochemical biosensor based on polyimide-boron nitride composite membranes. International Journal of Polymeric Materials and Polymeric Biomaterials. 2017; 66. p. 203-212. DOI: 10.1080/00914037.2016.1201763

Chapter 5

Polyimides for Micro-electronics Applications

Masao Tomikawa

Abstract

Polyimide is an organic polymer that exhibits the highest level of heat resistance, exhibits excellent mechanical properties and electrical insulation, and is stable for a long period of time. In addition, since it is easy to obtain polymers with different physical characteristics by changing the combination of monomers, it is possible to obtain the desired properties according to the application, and it is used in a wide range of fields such as insulating protective films for semiconductors and electronic components. This chapter describes polyimides used in microelectronics applications such as semiconductors, electronic components, displays, image sensors, and lithium-ion secondary batteries. The development of practical aspects such as photosensitivity, low-temperature curability, and adhesion to copper when used in microelectronics will be described.

Keywords: photosensitive polyimide, stress buffer, redistribution layer, poly(benzoxazole), refractive index, alignment film, pixel divided layer, planarization layer, binder resin, separator

1. Introduction (History to apply polyimide for micro-electronics)

Polyimide exhibits the highest level of heat resistance, excellent mechanical properties, and electrical insulation among organic resins, and therefore exhibits high reliability with little change in physical properties over a long period of time. Further, in general, a polyimide is obtained by reacting an acid anhydride and a diamine in a polar solvent to obtain a poly(amic acid) (PAA) as a polyimide precursor, and then converting to an inert polyimide by heat treatment [1].

Since the PAA is dissolved in a solvent or an alkaline aqueous solution, a polyimide pattern can be obtained in the state of the precursor by using such as photolithography technique.

Sato et al. examined the use of polyimide as an interlayer dielectrics for Integrated Circuit (IC) and showed that it performed higher reliability than the commonly used silicon dioxide [2]. Based on this result, polyimide has come to be used as an insulator of electronic components. Polyimides for those electronic applications required to form a pattern to make a circuit. Polyimide pattern was obtained by wet-etching process using hydrazine [3]. However, due to the toxicity of hydrazine, this method was abolished. Then partial imidized PAA was etched by tetra-methyl ammonium aqueous solution (TMAH), which is a developer of a positive photoresist. Etching of polyimide is performed at the same time as developing the photoresist [4].

In addition, May et al. found that memory data of DRAM was broken by α rays emitted from radioactive atoms contained as impurities in the ceramic package. The issue was called "soft error" [5]. They suggested to coat pure resin on the memory cell to absorb the alpha particle. After their report, polyimide coat that high purity resin was coated on a memory cell were effective to protect the soft error.

From the viewpoint of preventing the soft error in DRAM, the coating poly-imide on the semiconductor surface has been promoted [6, 7]. Further, it has been promoted to change the ceramic package to an epoxy mold resin composed to reduce a cost. The issue here is the thermal stress caused by the difference between the coefficient of thermal expansion of the semiconductor chip and the that of the epoxy mold resin. This thermal stress occurs during soldering and causes problems such as cracks in the passivation layer and epoxy mold resin, and deformation of aluminum wiring. To solve the issue, it has been proposed to apply a heat-resistant and flexible polyimide to the surface of the semiconductor chip. The polyimide layer to reduce thermal stress is called "stress buffer". The stress buffer is the main application of polyimide coating for semiconductor devices. The polyimide for the stress buffer is required to show good adhesion to Si, mold resin and metal with rather low modulus and good thermal stability [6, 7].

To form the stress buffer, a non-photosensitive polyimide was coated and etched by alkaline solution through a photoresist as a mask. Semiconductor manufactur-ers have been requested photosensitive polyimide which has a capability to make a lithographic patten by photolithographic technique, because of the high precision of pattering dimensions and the reducing pattering processes [8].

2. Negative tone photosensitive polyimide

Photosensitive polyimide was made by introducing a photosensitive group into polyimide or its precursor, or by adding a photosensitive component. The first reported photosensitive polyimide was the mixture of the dichromate compound and the PAA by Kerwin et al. [9]. Dichromate photo-resist which is composed of dichromate and water soluble resin such as PVA, casein, gelatin etc. is used in etching mask for lead frame [10]. Reaction mechanism of the dichromate photo resist is photo induced reduction of dichromium salt [11]. However, this method has not been used because it uses a highly toxic chromium compound and the solution stability is poor.

Rubner et al. synthesized a poly(amic ester) in which a photopolymerizable acrylic group alcohol was introduced into the carboxyl group of the PAA by an ester bond and obtained negative working photosensitive polyimide [12]. This method requires acid chloride in the reaction of dicarboxylic acid and diamine, which has drawbacks such as complicated synthesis process, removal of impurities, and dif-ficulty in removing photosensitive components during thermosetting. However, the technology was transferred to a polyimide manufacturer, and as results of vigorous research, it was widely applied in semiconductor stress buffer [13].

As a technology to counter the ester type, Hiramoto et al. have developed a simple negative photosensitive polyimide called "ionic bonded type". The photo-sensitive polyimide is compsed of PAA, tertiary amine having photo reactive group such as acrylic group [14]. The photo reacitve group was introduced PAA by ionic interaction between carboxylic acid and tertiary amine. This method is extremely easy to obtain photosensitive polyimide, and not only the photosensitive component easily volatilizes during thermosetting, but it also acts as a catalyst for imidization, and curing is completed at a lower temperature. The ionic photosensitive polyimide was first practical use as an interlayer dielecrics for mounting substrates of super compters [15]. In addition, this photosensitive mechanism was unclear because no reaction of photoreactive acrylic groups was observed, but it was found that PAA causes photocharge separation with ultraviolet rays and the reaction proceeds [16].

Another negative type is a soluble polyimide composed of a benzophenone tetracarboxylic acid and a diamine having an alkyl group at the ortho position which was developed by Pfifer et al. [17]. The photosensitive mechanism of this polyimide was investigated by Horie et al., And the reaction mechanism was shown in which benzophenone was excited by ultraviolet rays (UV) to cause hydrogen abstraction from the alkyl group to crosslink and insolubilize it [18].

Furthermore, Omote et al. provided a negative-type image by adding a nifedipine to PAA [19]. The nifedipine changes its chemical structure and basicity by UV exposure. The interaction between PAA and nifedipine was changed by UV exposure due to basicity change of the nifedipine [19].

3. Positive tone photosensitive polyimide

There are two types of photoresists, a negative type and positive type. Exposed area of the negative type resist came to insoluble by UV induced chemical reaction. On the other hand, that of positive photoresist came to soluble to soluble to alkaline solution by UV reaction. The negative type was first put into practical use, and then the positive type came out. Generally, in the negative type, an UV reactive group such as an acrylic group photopolymerizes with UV to form a crosslinked structure, so that the negative type is insoluble in a developing solution. It is difficult to form a fine pattern because the cross-linked polymer swells in the developer.

On the other hand, common positive photoresist is composed of novolak resin having a phenolic hydroxyl group soluble in an alkaline aqueous solution, and diazonaphthoquinone compound. The diazonaphthoquinone compound is insoluble in alkali and formed a complex with novolak resin before UV exposure. The diazonaphthoquinone compound converts to indencarboxylic acid by UV exposure [20]. The indenecarboxylic acid is alkaline soluble. As a result, uncxposed area is hard to soluble to alkaline solution and exposed area is soluble to alkaline solution [21]. The development of polyimide of this technology was also studied from an early stage, and Loprest et al. invented a positive photosensitive polyimide precursor using a PAA and a diazonaphthoquinone compound [22]. However, a good image cannot be obtained because the solubility of PAA to an alkaline aqueous solution is too large, and it has not been put into practical use as it is. The flow of this technology was subsequently announced by adding a diazonaphthoquinone compound to a polyimide or poly(amic acid) ester having a phenolic hydroxyl group [23–25]. In addition, Tomikawa et al. developed a partial esterification of PAA by using dimethylformamide dialkyl acetal. And the reaction made it possible to control the dissolution rate of the partial esterified PAA to alkaline solution [26].

Rubner et al. developed a positive heat-resistant material using polybenzoxazole (PBO) precursor as a heterocyclic polymer having heat resistance comparable to that of polyimide [27]. The precursor of PBO is polyhydroxyamide (PHA), which is a polyamide having a phenolic hydroxyl group, and has an appropriate alkali solubility. By adding a diazonaphthoquinone diazide compound to PHA, positive image of PBO was obtained. This technology has been deployed to various companies such as Sumitomo Bakelite and is widely used [28].

In addition, the development of positive photosensitive polyimide was also considered from another point of view. Kubota et al. announced a product using o-nitrobenzyl ester of PAA [29]. This is because the o-nitrobenzyl group is eliminated by deep UV exposure, so that the exposed part becomes PAA and becomes alkali-soluble. Furthermore, in the case of adding a nifedipine compound to the PAA developed by Omote et al. explained in the negative type, the hydrogen bonding strength changes depending on the baking conditions after exposure, and a positive type image can be obtained by controlling baking after exposure [30].

In addition, Tamura et al. found that a positive image can be obtained by baking an ion-bonded photosensitive polyimide at 130–150° C after exposure [31]. Regarding this mechanism, the glass transition temperature (Tg) differs between the exposed area and the unexposed area, and the Tg of the exposed area is slightly higher compare to that of unexposed area. When baking is performed near Tg, imidization rate of exposed area is slower than that of unexposed area due to Tg difference. High Tg, imidization of the exposed area with high Tg does not proceed, and imidization of the unexposed part with low Tg progresses, so that the exposed area with high Tg becomes an alkaline developer. It was found that a positive image was obtained [32].

The mainstream of current photoresists is chemical amplification type which is composed of alkali soluble resin protected by acid cleavable group and photo acid generator (PAG). An acid is generated from PAG by UV exposure and de-protection of acid cleavable group proceeds. Then the exposed area became alkaline soluble. This technology was also applied for polyimide. So solvent-soluble polyimide having a phenolic hydroxyl group, which was protected by an acid-cleavable protective group such as a t-BOC (tert-butoxycarbonyl) group and photo acid generator (PAG) [33, 34]. Nakano et al. developed the composition of PAA oligomer and a methylol compound are crosslinked at the time of prebaking and cleaved with an acid generated from PAG after UV exposure to obtain a positive image [35].

Furthermore, Ueda et al. proposed a ternary system using a acid cleavable dissolution inhibitor and a PAG to enhance dissolution contrast between exposed and unexposed areas and to use polyimide and PBO precursors as they are [36]. According to this method, it is not necessary to protect the polymer itself, and by adding a dissolution inhibitor with a protecting group that is eliminated by acid, the dissolution rate ratio between the exposed and unexposed areas may exceed 2000.

Ohyama obtained positive photosensitive polyimide by reaction development patterning [37]. In addition, they are obtained positive photosensitive polyimide from polyerimide (Ultem), which is alkali insoluble thermoplastic polyimide, a diazonaphthoquinone compound. They designed a development solution also by mixing aqueous TMAH aqueous solution, a NMP as a solvent and a nucleophilic base such as monoethanolamine to proceed decomposition. As a result, they obtained a positive polyetherimide image [38]. Furthermore, it was shown that engineering plastics such as polycarbonate can be used in this method [39]. They investigated and stimulated development mechanism and found that a salt composed of an acid made of a diazonaphthoquinone compound and an alkali of the developer accelerates the penetration of the hydrophilic developer into the exposed area. The nucleophilic reaction in the exposed area proceeds, and the main chain is decomposed. On the other hand, it has been reported that the unexposed area does not form salt in the developing solution and the reaction of the main chain is negligible, so that the unexposed area remains [37]. This technique can also obtain negative images, which was obtained by adding phenylmaleimide and diazonaphthoquinonediazide compounds to the polymer, and developing with a developer containing alcohol in a TMAH aqueous solution [40]. Furthermore, it has been reported that development with an aqueous solution of TMAH, which is generally used in semiconductor processing, is also possible [39]. This method is an interesting technique because it shows that a polymer having more excellent physical properties can be applied as a photosensitive materials. It has also been reported that even when a polyisoimide and a diazonaphthoquinone compound are added, the exposed portion becomes alkali-soluble and an image is obtained [41].

4. Low temperature curable photosensitive polyimide

New photosensitive polyimide applications have emerged for semiconductor packages in addition of stress buffer. As semiconductors and electronic components are becoming smaller, the boards on which they are mounted are also becoming smaller, and the components are placed on the board and heat-mounted from the method of inserting the pins as electrodes into the holes in the printed circuit board (PCB). Surface mount technology (SMT) has been developed to minimize the footprint of semiconductors and electronic components [42]. In SMT, the substrate on which semiconductors and electronic components are mounted is raised to a temperature at solder melting, and mounting is performed. Therefore, in semiconductors with not so many electrodes, chip scale packages (CSPs) have emerged in which the electrodes are formed into convex bumps at the bottom rather than around the semiconductor [43]. In addition, for those with many lead-out electrodes, bump formation is not sufficient with the semiconductor package alone, and a fan-out type package (FO-WLP) that enables bump formation even in the mold resin portion has appeared [44]. For both CSP and FO-WLP, a re-distribution layer (RDL) is formed using photosensitive polyimide or PBO so that bumps can be formed using the entire surface of the semiconductor package [45, 46].

To make a FO-WLP, RDL is formed outside the chip (**Figure 1**). Therefore, the chip is put into the mold resin, and rewiring is also formed in the mold resin portion. Therefore, the material used for rewiring needs to be formed below the heat resistant temperature of the mold resin composed of epoxy resin and silica filler.

As a result, materials that can be fired at 200° C or lower are required for rewiring applications. Furthermore, in FO-WLP, since photosensitive polyimide and PBO are directly bonded to LSI and mold resin, it is necessary that the photosensitive polyimide, PBO and solder bumps are not destroyed by thermal shock due to the difference in thermal expansion coefficient. Mechanical properties such as elongation at break are regarded as important, and product development is being carried out for this purpose [47]. In addition, FO-WLP is becoming finer, and wiring is being formed with a line and space of 2 μm [48].

Various studies have been conducted for low temperature curing. In order to perform imide ring closure at a low temperature, a thermobase generator is added [49], and an pre-imidized polymer is used [47]. In addition, focusing on the main chain structure of polyimide, Sasaki reported that the imidization rate determined by the acidity of the diamine component. So diamine having high acidity gave low temperature imidization [50].

Furthermore, PBO is also being studied for low temperature cyclization. If the structure constituting PBO is made flexible, low temperature curing is possible [51], and a thermoacid generator that generates sulfonic acid by heat is added. It has been reported that it cyclizes at low temperatures [52]. Furthermore, Kusunoki

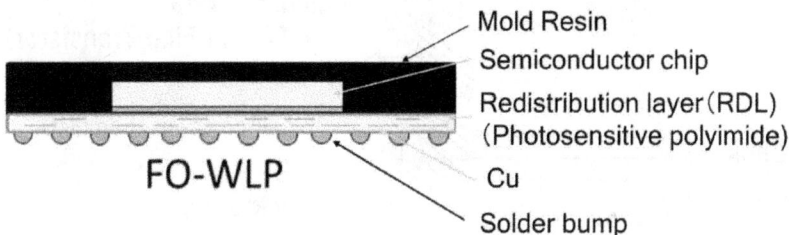

Figure 1.
Cross section structure of FO-WLP.

reported photosensitive polynorbornene which has low modulus, low dielectric constant, low water absorption with low temperature curable [53].

5. Polyimide for image sensor

As the image sensor becomes smaller and more integrated, each pixel becomes smaller, the amount of incoming light decreases, and the sensitivity decreases. On the other hand, a micro-lens is formed on the pixel to increase the light flux. Suwa et al. dispersed titania sol in a positive photosensitive polyimide to obtain a positive photosensitive heat-resistant resin having a refractive index of 1.9. This material is further made lenticular in the curing process [54]. In addition, they developed a siloxane base photosensitive high refractive index material whose refractive index is 1.9 [55]. On the other hand, Oishi et al. focused on the triazine skeleton, and obtained a material having a refractive index of 1.80 by using a hyperbranched polymer of triazine without adding a filler having a high refractive index [56].

6. Polyimide for LCD

One of big applications of polyimides to a display is an alignment layer of a liquid crystal display (LCD). The purpose of alignment film is arranging liquid crystal molecules in LCD (**Figure 2**). The alignment film is rubbed by a cloth to align liquid crystal molecules. Although the detailed mechanism by which the rubbing-treated polyimide orients the liquid crystal molecules is unknown, it was found that the polyimide is suitable as a liquid crystal alignment film, and it has come to be used in liquid crystal displays. Initially, aromatic polyimides such as PMDA and ODA were used for the alignment film of TN (Twisted Nematic) type LCD. Then STN (Super Twisted Nematic) type LCD was developed. New polyimide was developed for STN LCD alignment film [57].

To widen the viewing angle VA (Vertical Alignment) type, and IPS (In Plane Switching) type LCDs were shown [58].

To improve the LCD, numerous studies have been conducted on the relationship between the alignment state of the liquid crystal and the alignment film from the viewpoint of the influence of the rubbing treatment and the molecular structure of polyimide [59–62]. Furthermore, after that, a series of systematic basic studies were conducted using synchrotron radiation equipment [63–66].

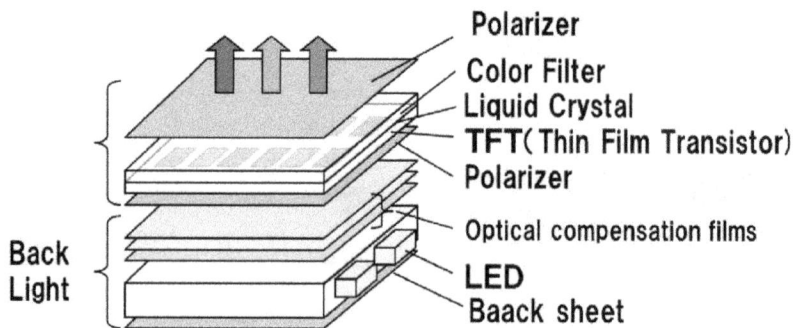

Figure 2.
Cross section structure of Liquid Crystal Display.

From those basic studies design guides for the alignment film were shown [67]. Various types of polyimide alignment materials were commercialized [68].

Since dust is generated in the rubbing process, a method of aligning the liquid crystal without rubbing has been studied [69]. Photo active polyimide were suggested for the phot rubbing process. Photo active polyimide means photo rearrangement type and photo decomposed type [69–72].

7. Polyimide for OLED

Compared to LCD, organic EL displays (OLED) are self-luminous, have a large contrast between emitting area and non-emitting area, have a simple layer structure and can be made thin, have a fast response speed, and can display even on flexible substrate. The cross-sectional structure of the OLED is shown in the **Figure 3**. There are electrodes above and below the emission layer, and the light emitting layers are separated from each other by a pixel divided layer (PDL). Since the PDL contact with the emission layer, The water and decomposed gas generated from the PDL damage the adjacent to emission layer, narrowing the light emitting region. Heat resistance and degassing properties are important. In addition, the partition wall is not like a rectangular parallelepiped, but a trapezoidal shape with a gentle taper is preferable to suppress the current concentration.

On the other hand, another organic resin in OLED is PNL (Planarization Layer). The PNL is akso required low out gas and moisture. Therefore, a positive photosensitive polyimide material is suitable as a PDL and PNL for OLED display and is widely used [73].

8. High frequency application

In the future, mobile phones will be able to send and receive large amounts of data in a short time by utilizing high frequencies, the operating frequency of application processors that control mobile phones to become multifunctional will be higher, and automobile collision safety. Due to the increasing adoption of millimeter-wave radar to improve performance, the use of high-frequency materials will increase more than ever. For those applications, materials such as fluoropolymer [74], liquid crystal polyester [75], BCB (benzocyclobutene) [76], polyphenylene ether [77], and cycloolefin polymer [78] have a dielectric constant in the high frequency region. It has been used because of its low dielectric loss, but there are also problems such as low adhesiveness, and low dielectric constant and

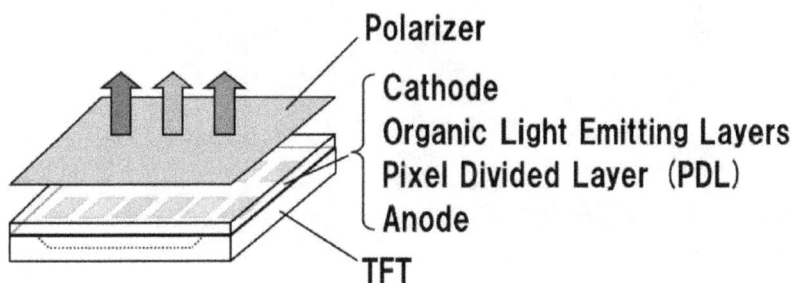

Figure 3.
Cross section structure of OLED display.

low dielectric loss are being studied by making porous polyimide [79]. Authors obtain a low dielectric loss photosensitive polyimide design guide line by dynamic mechanical analysis [80].

9. Li ion battery

Lithium-ion secondary batteries (LiB) are used widely as high energy density storage batteries for mobile phones, notebook PCs, electric tools, electric vehicle and so on, due to their high discharge voltage, almost no memory effect of charging and discharging, and small self-discharge. The battery is composed of negative electrode (Anode), positive electrode (Cathode), separator, and electrolytic solution as shown in **Figure 4**. Anode is composed of anode active material, binder resin, and Cu foil. Cathode is composed of cathode active material, and Al foil. SBR (styrene butadiene rubber), PVDF (Polyvinylidene fluoride) are used for binder resin. Conductive assisting agent is added to decrease electric resistivity in binder resin. Current anode acting material is carbon-based materials such as graphite or hard carbon. Lithium ions move in and out between the layers of graphite to charge and discharge, and the theoretical capacity is 372mAh / g. Recent batteries have almost reached this theoretical value, and it is necessary to change the active material to further increase the capacity. Silicon is attracting as an anode active material candidate for next generation due to its large capacity [81]. However, the volume changes of the Si active material is about 400% at charging / discharging. To use large volume change Si material, new binder resin is required to endure the big volume change. Promising candidate is polyimide which has excellent mechanical properties and good adhesive properties. It has been reported that the cycle characteristics are significantly improved when a polyimide-based material is used as a binder for such a Si base large-capacity anode active material [82–84].

On the other hand, Li metal oxide base material is used for cathode active compound. Common binder resin for cathode active material is PVDF. To improve the safety, olivin compound such as LiFePO4 (LFO) is developed [85]. Li ion capacity of LFO is small compare to Li metal oxide material such as LiCoO2, LiNiO3. But LFO shows higher thermal stability than those of Li metal oxides. So operation temperature of LFO is wider than that of Li metal oxide. Miyuki et al. reported that polyimide is good binder for LFO due to high temperature operation, high charging / discharging property [86].

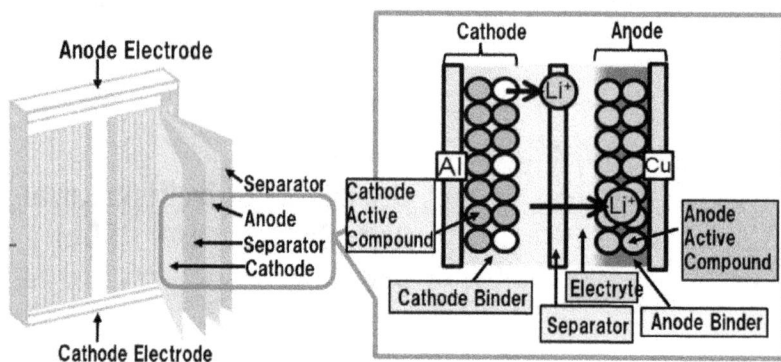

Figure 4.
Schematic diagram of lithium ion secondary battery.

In addition, porous polyimide separator was obtained from polyimide and silica filler. The separator shows high rate and long term stability [87].

10. Conclusion

It has been more than 50 years since polyimide was put into practical use, but development for new applications is still underway. This is due to the high degree of freedom in design that allows the material called polyimide to be easily synthesized, maintains excellent heat resistance, and changes other physical properties in various ways. It is expected that new developments will continue in the future due to the degree of freedom in design and excellent characteristics such as heat resistance and insulation.

Author details

Masao Tomikawa
Research and Development Division, Toray Industries Inc., Otsu, Shiga, Japan

*Address all correspondence to: masao.tomikawa.j4@mail.toray

IntechOpen

References

[1] C. E. Sroog, A.L. Endrey, S.V. Abramo, C.E. Berr, W.M. Edward, and K.L. Oliver; J. Polym. Sci., Part A, 3, 1373-1390 (1965), DOI10.1002/pol.1965.100030410

[2] K. Sato, S. Harada, A. Saiki, T. Kimura, T. Okubo and K. Mukai, "A Novel Planar Multilevel Interconnection Technology Utilizing Polyimide", IEEE Trans. Parts Hybrid and Packaging, vol. PHP-9, pp. 176-180, 1973. DOI: 10.1109/TPHP.1973.1136727

[3] J. I. Jones, "The Reaction of Hydrazine with Polyimides, and Its Utility2. J. Polym. Sci., C Polym. Symp., 22, 773-784 (1969), https://doi.org/10.1002/polc.5070220219

[4] G.C. Davis, and C.L. Fasoldt; "Wet Etch Patterning of Polyimide Siloxane For Electronic Applications", Proc. 2nd Ellenville Conf. on Polyimides, 381 (1987).

[5] May, T. and M. H. Woods. "A New Physical Mechanism for Soft Errors in Dynamic Memories." 16th International Reliability Physics Symposium, pp. 33-40, (1978), DOI:10.1109/IRPS.1978.362815

[6] R. C. Baumann, "Investigation of the Effectiveness of Polyimide Films for the Stopping of Alpha Particles in Megabit Memory Devices" TI. Tech. Report, April 1991.

[7] K. Kitade, S. Koyama, N. Motoki, K. Mitsusada and K. Asakura, Prevention of Soft Error by Polyimide Resin Coating, Proc. IECE of Japan Semiconductor and Materials, pp. 63, 1981.

[8] C. Schuckert, D. Murray, C. Roberts, G. Cheek and T. Goida, "Polyimide stress buffers in IC technology," IEEE/SEMI Conference on Advanced Semiconductor Manufacturing Workshop, pp. 72-74 (1990) doi: 10.1109/ASMC.1990.111222.

[9] R. E. Kerwin, and M. R. Goldrick, "Thermally stable photoresist polymer", Polym. Eng. Sci, 11: 426-430 (1971), https://doi.org/10.1002/pen.760110513

[10] R. Ueda,"Chemical Machining by Ferric Chloride Etchant", Corr. Eng., 38(4), 231-237 (1989) https://doi.org/10.3323/jcorr1974.38.4_231

[11] L. Grimm, K. J. Hilke and E. Scharrer, "The Mechanism of the Cross Linking of Poly (Vinyl Alcohol) by Ammonium Dichromate with UV-Light", J. Electrochem. Soc., 130(8), 1767-1771 (1983). https://doi.org/10.1149/1.2120089

[12] R. Rubner, B. Bartel and G. Bald;" Production of Highly Heat-Resistant Film Patterns from Photoreactive Polymer Precursors. Part I. General Principles", Siemens Forsch Entwickl. Ber., 5, 235-239 (1976).

[13] C. Schuckert, D. Murray, C. Roberts, G. Cheek, and T. Goida, IEEE/SEMI Advanced Semiconductor Manufacturing Conference, 72-74 (1990).

[14] N. Yoda and H. Hiramoto, "New Photosensitive High Temperature Polymers for Electronic Application", J. Makromol. Sci. Chem., A21, 1641-1663 (1984), https://doi.org/10.1080/00222338408082082

[15] T. Ohsaki, T. Yasuda, S. Yamaguchi and T. Kon, "A Fine-Line Multilayer Substrate with Photo Sensitive Polyimide Dielectric and Electroless Copper Plated Conductors", Proc. 3rd IEMT Symp., pp. 178-183, 1987.

[16] M. Tomikawa, M. Asano, G. Ohbayashi, H. Hiramoto, Y. Morishima and M. Kamachi; "Photo-reaction of

Ionic bonding Photosensitive Polyimide", J. Photo Polym. Sci. & Technol., 5, 343-350 (1992).

[17] J. Pfeifer and O. Rohde, in Proceedings of the 2nd International Conference on Polyimides, Society of Plastics Engineers Inc., Ellenville, NY, 1985, pp. 130-151.

[18] H. Higuchi, T. Yamashita, K. Horie and I. Mita, "Photo-Cross-Linking Reaction of Benzophenone-Containing Polyimide and Its Model Compounds", Chem. Mater. 3, 188-194 (1991) DOI: 10.1021/cm00013a038

[19] T. Omote and T. Yamaoka; "A new positive-type photoreactive polyimide precursor using 1,4-dihydropyridine derivative".Polym. Eng. Sci., 32, 1634-1641 (1992), https://doi.org/10.1002/pen.760322117

[20] O. Süs, "Über die Natur der Belichtungsprodukte von Diazoverbindungen. Übergänge von aromatischen 6-Ringen in 5-Ringe. Justus Liebigs Ann. Chem., 556, 65-84. (1944) https://doi.org/10.1002/jlac.19445560107

[21] M. Hanabatake and A. Furuta," Applications of High-ortho Novolak Resins to Photoresist Materials", Kobunshi Ronbunshu, 45(10), 803-808 (1989), https://doi.org/10.1295/koron.45.803

[22] Frank J. Loprest Eugene F. McInerney," Positive working thermally stable photoresist composition, article and method of using", U.S. Pat. 4093461 (1978).

[23] Werner H. Mueller, and Dinesh N. Khanna, "Hydroxy polyimides and high temperature positive photoresists therefrom", US patent 4927736, 1990.

[24] Khanna, D.N. and Mueller, W.H. "New high temperature stable positive photoresists based on hydroxy

polyimides and polyamides containing the hexafluoroisopropylidene (6-f) linking group", Polym Eng Sci, 29: 954-959, (1989), https://doi.org/10.1002/pen.760291414

[25] S.L.-C. Hsu, P.-I. Lee, J.-S. King, and J.-L. Jeng, "Novel positive-working aqueous-base developable photosensitive polyimide precursors based on diazonaphthoquinone-capped polyamic esters", J. Appl. Polym. Sci., 90, 2293-2300 (2003), https://doi.org/10.1002/app.12905

[26] 26.Tomikawa, M., Yoshida, S. & Okamoto, N. Novel Partial Esterification Reaction in Poly(amic acid) and Its Application for Positive-Tone Photosensitive Polyimide Precursor. Polym J 41, 604-608 (2009). https://doi.org/10.1295/polymj.PJ2008333

[27] Rubner, R. (1990), Photoreactive polymers for electronics. Adv. Mater., 2: 452-457. https://doi.org/10.1002/adma.19900021003

[28] H. Makabe, T. Banba, and T. Hirano; "A Novel Positive Working Photosensitive Polymer For Semiconductor Surface Coating", J. Photopolym. Sci. & Technol., 10, 307-311 (1997), https://doi.org/10.2494/photopolymer.10.307

[29] S. Kubota, Y. Tanaka, T. Moriwaki, and S. Eto, "Positive Working Photosensitive Polyimide: The Effect of Some Properties on Sensitivity", J. Electochem. Soc.,138, 1080-1084 (1991), https://doi.org/10.1149/1.2085719

[30] T. Yamaoka, S. Yokoyama, T. Omote, K. Naito, and K. Yoshida; "Photochemical Behavior of Nifedipine Derivatives and Application to Photosensitive Polyimides", J. Photopolym. Sci. & Technol., 9, 293-304 (1996),, https://doi.org/10.2494/photopolymer.9.293

[31] K. Tamura, M. Eguchi, and M. Asano, "Positive Photosensitive Polyimide Precursor Comosition", JP H06-324493, 1994,,

[32] S. Yoshida, M. Eguchi, K. Tamura, and M. Tomikawa, "Positive Pattern Formation and its Mechanism from Ionic bonded Negative Photosensitive Polyimide", J. Photopolym. Sci. & Technol., 20, 145-147 (2007), https://doi.org/10.2494/photopolymer.20.145

[33] T, Omote, Koseki, and T, Yamamoka, "Fluorine-containing photoreactive polyimide. 6. Synthesis and properties of a novel photoreactive polyimide based on photo-induced acidolysis and the kinetics for its acidolysis", Macromol., 23, 4788-4795 (1990), https://doi.org/10.1021/ma00224a007

[34] R. Hayase, N. Kihara, N. Oyasato, S. Matake, and M. Oba, "Positive photosensitive polyimides using polyamic acid esters with phenol moieties". J. Appl. Polym. Sci., 51: 1971-1978 (1994) https://doi.org/10.1002/app.1994.070511113

[35] T. Nakano, H. Iwasawa, N. Miyazawa, S. Takahara, and T. Yamamoka, "Positive-Type Photopolyimide Based on Vinyl Ether Crosslinking and De-Crosslinking", J. Photopolym. Sci. & Technol., 13, 715-718 (2000), https://doi.org/10.2494/photopolymer.13.715

[36] T. Ogura, T. Higashihara and M. Ueda; "Development of Photosensitive Poly(benzoxazole) Based on a Poly(o-hydroxy amide), a Dissolution Inhibitor, and a Photoacid Generator", J. Photopolym. Sci. & Technol.,22, 429-435 (2009), https://doi.org/10.2494/photopolymer.22.429

[37] T. Fukushima, T. Oyama, T. Iijima, M. Tomoi, and H. Itatani," New concept of positive photosensitive polyimide: Reaction development patterning

(RDP)", J. Polym. Sci. A Polym. Chem., 39, 3451-3463 (2001), https://doi.org/10.1002/pola.1327

[38] T. Fukushima, Y. Kawakami, T. Oyama, and M. Tomoi, "Photosensitive Polyetherimide (Ultem) Based on Reaction Development Patterning (RDP)", J. Photopolym. Sci. & Technol., 15, 191-196 (2002), https://doi.org/10.2494/photopolymer.15.191

[39] T. OYAMA, "Novel Technique fbr Changing Engineering Plastics to Photosensitive Polymers", Kobunshi, 55, 887 (2006), https://doi.org/10.1295/kobunshi.55.887, https://www.jstage.jst.go.jp/article/kobunshi1952/55/11/55_11_887/_article/−char/en

[40] Oyama T, Sugawara S, Shimizu Y, Cheng X, Tomoi M, Takahashi A. "A Novel Mechanism to Afford Photosensitivity to Unfunctionalized Polyimides: Negative-Tone Reaction Development Patterning". J Photopolym. Sci. Technol. 2009;22: 597-602.

[41] 41.A. Mochizuki, T. Teranishi, M. Ueda, K. Matsushita, "Positive-working alkaline- developable photosensitive polyimide precursor based on polyisoimide using diazonaphtho quinone as a dissolution inhibitor", Polymer 36 (11) (1995) 2153-2158.

[42] Y. Satoh, T. Samejima, "Surface Mount Technology of Chip Components", Circuit Technology, 3(3), 188-197 (1988), DOI: https://doi.org/10.5104/jiep1986.3.188,

[43] M. Yasunaga, S. Baba, M. Matsuo, H. Matsushima, S. Nakao and T. Tachikawa, "Chip scale package: A lightly dressed LSI chip", IEEE Trans. Comp. Packag. Manufact. Technol., vol. 18, pp. 451-457, 1995. DOI: 10.1109/95.465135

[44] M. Brunbauer, E. Fergut, G. Beer, T. Meyeer, H. Hedler, J. Belonio, E.

Nomura, K. Kikuchi, K. Kobayashi, "An embedded device technology based on a molded reconfigured wafer", in Proc. 56th Electron. Compon. Technol. Conf., May/Jun. 2006, pp. 547-551, DOI: 10.1109/ECTC.2006.1645702

[45] T. Yuba, M. Suwa, Y. Fujita, M. Tomikawa, and G. Ohbayashi; "A Novel Positive Working Photosensitive Polyimide For Wafer-level CSP Packages", J. Photopolym. Sci. & Technol., 15, 201-203 (2002) DOI https://doi.org/10.2494/photopolymer. 15.201

[46] K. Yamamoto, and T. Hirano; "Application for WLP at positive working photosensitive polybenzoxazole", J. Photopolym. Sci. & Technol., 15, 173-176 (2002) DOI https://doi.org/10.2494/photopolymer. 15.173

[47] K. Fukukawa, T. Ogura, Y. Shibasaki, and M. Ueda; Chem. Lett., 34, 1372 (2005), "Thermo-base Generator for Low Temperature Solid-phase Imidation of Poly(amic acid)", Chem. Let. 2005 34(10), 1372-1373 DOI https://doi.org/10.1246/cl.2005.1372

[48] Won Kyoung Choi, Duk Ju Na, Kyaw Oo Aung, Andy Yong, Jaesik Lee, Urmi Ray, Riko Radojcic, Bernard Adams, Seung Wook Yoon; "Ultra Fine Pitch RDL Development in Multi-layer eWLB (embedded Wafer Level BGA) Packages". International Symposium on Microelectronics 1 October 2015; 2015 (1): 822-826. doi: https://doi. org/10.4071/isom-2015-THP34

[49] Y. Shoji, Y. Koyama, Y. Masuda, K. Hashimoto, K. Isobe, and R. Okuda; "Development of Novel Low-temperature Curable Positive-Tone Photosensitive Polyimide with High Elongation", J. Photopolym. Sci & Technol., 29, 277 (2016) DOI https://doi. org/10.2494/photopolymer.29.277

[50] H. Onishi, S. Kamemoto, T. Yuba and M. Tomikawa; "Low-temperature Curable Positive-tone Photosensitive Polyimide coatings", J. Photopolym. Sci. & Technol., 25, 341-344 (2012) DOI https://doi.org/10.2494/photopolymer. 25.341

[51] T. Sasaki; "Low Temperature Curable Polyimide for Advanced Package", J. Photopolym. Sci & Technol., 29, 379-382 (2016) DOI https://doi.org/10.2494/photopolymer. 29.379

[52] K. Iwashita, T. Hattori, S. Ando, F. Toyokawa, and M. Ueda, "Study of Polybenzoxazole Precursors for Low Temperature Curing",. J. Photopolym. Sci. & Technol., 19, 281-282 (2006) DOI https://doi.org/10.2494/photopolymer. 19.281

[53] F. Toyokawa, Y. Shibasaki, and M. Ueda; "A Novel Low Temperature Curable Photosensitive Polybenzoxazole", Polym. J., 37, 517-521 (2005) DOI https://doi.org/10.1295/ polymj.37.517

[54] M. Suwa,. H Niwa, and M. Tomikawa, "High Refractive Index Positive Tone Photo-sensitive Coating", J. Photopolym. Sci. & Technol., 2006, Volume 19, Issue 2, Pages 275-276, Released August 15, 2006, Online ISSN 1349-6336, Print ISSN 0914-9244, https://doi.org/10.2494/photopolymer. 19.275,

[55] T. Hibino, M. Naruto, Y. Imanishi, and M. Suwa, "High Refractive Index Photo-sensitive Siloxane Coatings", J. Photopolym. Sci. & Technol., 2019, 32 (3) p. 485-488, 公開日 2019/11/14, Online ISSN 1349-6336, Print ISSN 0914-9244, https://doi.org/10.2494/ photopolymer.32.485, https://www. jstage.jst.go.jp/article/photopolymer/ 32/3/32_485/_article/-char/ja, 抄録:

[56] N. Nishimura, Y. Shibasaki, M. Ozawa, and Y. Oishi, "High Refractive Coating Materials Using Hyperbranched Polymers", J. Photopolym. Sci. &

Technol., 2012, Volume 25, Issue 3, Pages 355-358, Released August 23, 2012, Online ISSN 1349-6336, Print ISSN 0914-9244, https://doi.org/10.2494/photopolymer.25.355,

[57] H. Fukuro and S. Kobayashi, "NEWLY SYNTHESIZED POLYIMIDE FOR ALIGNING NEMATIC LIQUID-CRYSTALS ACCOMPANYING HIGH PRETILT ANGLES", Mol. Cryst. Liq. Cryst. 163, pp. 157-162 (1988) DOI10.1080/00268948808081995

[58] K. Ono, "Progress of an IPS Panel Technology for LCD-TVs", J. the Imaging Soc., Jpn, 47(5), pp. 433-438 (2008), https://doi.org/10.11370/isj.47.433

[59] K. Sawa, K. Sumiyoshi, Y. Hirai, K. Tateishi, and T. Kamejima; "Molecular orientation of polyimide films for liquid crystal alignment studied by infrared dichroism", Jpn. J. Appl. Phys., 33, 6273-6276 (1994). DOI: 10.1143/JJAP.33.6273

[60] H. Nejoh; "Liquid crystal molecule orientation on a polyimide surface, Surf. Sci., 256 (1-2), pp. 94-101 (1991), ISSN 0039-6028, https://doi.org/10.1016/0039-6028(91)91203-A.

[61] N, A.J. M van Aerle, and A.J.W. Tel; "Molecular Orientation in Rubbed Polyimide Alignment Layers Used for Liquid-Crystal Displays", Macromol., 27, 6520-6526 (1994) https://doi.org/10.1021/ma00100a042

[62] K. Weiss, C. Woll. E. Bohm, B. Fiebranz, G. Forstmann, B. Peng, V. Scheumann, and D. Johannsmann; "Molecular Orientation at Rubbed Polyimide Surfaces Determined with X-ray Absorption Spectroscopy: Relevance for Liquid Crystal Alignment", Macromol., 31, 1930-1936 (1998) https://doi.org/10.1021/ma971075z

[63] J. Stohr, and MG. Samant; "Liquid crystal alignment by rubbed polymer surfaces: a microscopic bond orientation model", J. Electron Spectrosc. Relat. Phenom, 98-99, 189 (1999).

[64] T. Koganezawa, I. Hirosawa, H. Ishii, and ; "Characterization of Liquid Crystal Alignment on Rubbed Polyimide Film by Grazing-Incidence X-Ray Diffraction", IEICE Trans. Electron. E92.C, 1371 (2009).

[65] I. Hirosawa, T Koganezawa, and H. Ishii; "Thickness of Crystalline Layer of Rubbed Polyimide Film Characterized by Grazing Incidence X-ray Diffractions with Multi Incident Angle, IEICE Transactions on Electron Devices, E97-C 11, 1089 (2014).

[66] N. Kawatsuki, Y. Inada, M. Kondo, Y. Haruyama, and S Matsui, "Molecular Orientation at the Near-Surface of Photoaligned Films Determined by NEXAFS", Macromol. 47, pp. 2080-2087 (2014), https://doi.org/10.1021/ma5000738

[67] M. Nishikawa, "Design of polyimides for liquid crystal alignment films", Polym. Adv. Technol., 11, 404-412. (2000) https://doi.org/10.1002/1099-1581(200008/12)11:8/12 < 404::AID-PAT41 > 3.0.CO;2-T.

[68] M. Nishikawa, "Development of Novel Polyimide Alignment Films for Liquid Crystal Display Televisions", J. Photopolym. Sci. & Technol., 24(3), p. 317-320 (2011) https://doi.org/10.2494/photopolymer.24.317,

[69] W. Gibbons, P. Shannon, ST. Sun, and B. J. Swetlin, "Surface-mediated alignment of nematic liquid crystals with polarized laser light", Nature 351, 49-50 (1991). https://doi.org/10.1038/351049a0

[70] K. Usami, "Influence of molecular structure on anisotropic photoinduced decomposition of polyimide molecules", J. Appl. Phys., 89, 5339-5342 (2001) https://doi.org/10.1063/1.1358323

[71] S. W. Lee, S. I Kim, B. Lee, H. C. Kim, T. Chang, and M Ree, "A Soluble Photoreactive Polyimide Bearing the Coumarin Chromophore in the Side Group: Photoreaction, Photoinduced Molecular Reorientation, and Liquid-Crystal Alignability in Thin Films", Langmuir 19(24), 10381-10389 (2003) https://doi.org/10.1021/la0348158

[72] K. Usami, K. Sakamoto, N. Tamura, and A. Sugimura, "Improvement in photo-alignment efficiency of azobenzene-containing polyimide films", Thin Solid Films, 518(2), 729-734 (2009) ISSN 0040-6090, https://doi.org/10.1016/j.tsf.2009.07.079.

[73] R. Okuda, K. Miyoshi, N. Arai and M. Tomikawa, "Polyimide Coatings for OLED Applications", J. Photopolym. Sci. & Technol., 17(2), 207-213 (2004) https://doi.org/10.2494/photopolymer.17.207,

[74] G. Hougham, G. Tesoro, and A. Viehbeck, "Influence of Free Volume Change on the Relative Permittivity and Refractive Index in Fluoropolyimides", Macromol., 29(10), 3453-3456 (1996) https://doi.org/10.1021/ma950342391.

[75] G. Zou, H. Gronqvist, J. P. Starski and J. Liu, "Characterization of liquid crystal polymer for high frequency system-in-a-package applications," IEEE Transactions on Advanced Packaging, vol. 25(4), 503-508 (2002), doi: 10.1109/TADVP.2002.807593.

[76] V. B. Krishnamurthy, H. S. Cole and T. Sitnik-Nieters, "Use of BCB in high frequency MCM interconnects," IEEE Transactions on Components, Packaging, and Manufacturing Technology: Part B, 19(1), 42-47 (1996), doi: 10.1109/96.486483.

[77] Y. Seike, Y. Okude, I. Iwakura, I. Chiba, T. Ikeno, and T. Yamada, "Synthesis of Polyphenylene Ether Derivatives: Estimation of Their Dielectric Constants". Macromol. Chem.

Phys., 204, 1876-1881 (2003), https://doi.org/10.1002/macp.200300002

[78] M. Yamazaki, "Industrialization and application development of cyclo-olefin polymer", J. Molecul. Cat.s A: Chem.,213(1), 81-87 (2004) ISSN 1381-1169,https://doi.org/10.1016/j.molcata.2003.10.058.

[79] Y. Ren, and D. Lam," Properties and Microstructures of Low-Temperature-Processable Ultralow-Dielectric Porous Polyimide Films". J. Elec Material., 37, 955-961 (2008). https://doi.org/10.1007/s11664-008-0446-z

[80] H. Araki, Y. Kiuchi, A. Shimada, H. Ogasawara, M. Jukei, and M. Tomikawa, "Low Df Polyimide with Photosensitivity for High Frequency Applications", J. Photopolym. Sci. & Technol., 33(2), 165-170 (2021), https://doi.org/10.2494/photopolymer.33.165,

[81] C.M. Park, J.H. Kim, H. Kim, H.J. Sohn, "Li-alloy based anode materials for Li secondary batteries", Chem. Soc. Rev., 39(8), 3115-3141 (2010), doi:10.1039/b919877f

[82] J. Oh, D. Jin, K. Kim, D. Song, Y. M. Lee, and M-H Ryou, "Improving the Cycling Performance of Lithium-Ion Battery Si/Graphite Anodes Using a Soluble Polyimide Binder", ACS Omega, 2 (11), 8438-8444 (2017), DOI: 10.1021/acsomega.7b01365

[83] J. S. Kim, W. Choi, K. Y. Cho, D. Byun, J. C. Lim, J. K. Lee, "Effect of polyimide binder on electrochemical characteristics of surface-modified silicon anode for lithium ion batteries", J. Power Sources, 244, 521-526 (2013), ISSN 0378-7753, https://doi.org/10.1016/j.jpowsour.2013.02.049.

[84] S. Uchida, M. Mihashi, M. Yamagata, and M. Ishikawa, "Electrochemical properties of non-nano-silicon negative electrodes prepared with a polyimide binder", J.

Power Sources, 273, 118-122 (2015),
ISSN 0378-7753, https://doi.
org/10.1016/j.jpowsour.2014.09.096.

[85] S-H. Wu, K. Hsiao K, and W-R. Liu,
"The preparation and characterization
of olivine LiFePO4 by a solution method
J. Power Sources, 146(1-2), 550-554
(2005), DOI 10.1016/j.jpowsour.2005.
03.128

[86] T. Miyuki, Y. Okuyama, T.
Sakamoto, Y. Eda, T. Kojima, and *T.
Sakai*, "Characterization of Heat
Treated SiO Powder and Development
of a LiFePO4/SiO Lithium Ion Battery
with High-Rate Capability and
Thermostability", Electrochem., 80(6),
401-404 (2012), DOI, https://doi.
org/10.5796/electrochemistry.80.401,

[87] Y. Shimizu, and K. Kanamura,
"Effect of Pore Size in Three
Dimensionally Ordered Macroporous
Polyimide Separator on Lithium
Deposition/Dissolution Behavior". J.
Electrochem. Soc., 166, A754– A761
(2019), DOI: 10.1149/2.1061904jes

www.ingramcontent.com/pod-product-compliance
Lightning Source LLC
Chambersburg PA
CBHW081231190326
41458CB00016B/5742